Lecture Notes in Networks and Systems 841

The series "Lecture Notes in Networks and Systems" publishes the latest developments in Networks and Systems—quickly, informally and with high quality. Original research reported in proceedings and post-proceedings represents the core of LNNS.

Volumes published in LNNS embrace all aspects and subfields of, as well as new challenges in, Networks and Systems.

The series contains proceedings and edited volumes in systems and networks, spanning the areas of Cyber-Physical Systems, Autonomous Systems, Sensor Networks, Control Systems, Energy Systems, Automotive Systems, Biological Systems, Vehicular Networking and Connected Vehicles, Aerospace Systems, Automation, Manufacturing, Smart Grids, Nonlinear Systems, Power Systems, Robotics, Social Systems, Economic Systems and other. Of particular value to both the contributors and the readership are the short publication timeframe and the world-wide distribution and exposure which enable both a wide and rapid dissemination of research output.

The series covers the theory, applications, and perspectives on the state of the art and future developments relevant to systems and networks, decision making, control, complex processes and related areas, as embedded in the fields of interdisciplinary and applied sciences, engineering, computer science, physics, economics, social, and life sciences, as well as the paradigms and methodologies behind them.

Indexed by SCOPUS, INSPEC, WTI Frankfurt eG, zbMATH, SCImago.

All books published in the series are submitted for consideration in Web of Science.

For proposals from Asia please contact Aninda Bose (aninda.bose@springer.com).

José Bravo · Gabriel Urzáiz
Editors

Proceedings of the 15th International Conference on Ubiquitous Computing & Ambient Intelligence (UCAmI 2023)

Volume 3

Editors
José Bravo
Ciudad Real, Spain

Gabriel Urzáiz
Mérida, Yucatán, Mexico

ISSN 2367-3370 ISSN 2367-3389 (electronic)
Lecture Notes in Networks and Systems
ISBN 978-3-031-48589-3 ISBN 978-3-031-48590-9 (eBook)
https://doi.org/10.1007/978-3-031-48590-9

This Springer imprint is published by the registered company Springer Nature Switzerland AG
The registered company address is: Gewerbestrasse 11, 6330 Cham, Switzerland

Paper in this product is recyclable.

Preface

Ubiquitous computing (UC) is a paradigm that allows software applications to obtain and process environmental information in order to make users feel that changes in the environment do not affect the functionality provided to them, thus aiming at making the technology invisible. UC is made possible by the confluence of computing, communication and control technologies, and it involves measuring and considering context variables (e.g., time and location of the users) to appropriately adapt the functionality of software applications to user needs.

On the other hand, ambient intelligence (AmI) represents functionality embedded into the environment that allows systematic and unattended sensing, as well as proactive acting, to provide smart services. Thus, AmI solutions frequently appear in domains of "being helped," e.g., social, psychological, healthcare and instrumental scenarios.

The UCAmI Conference presents advances in both AmI application scenarios and their corresponding technical support through UC. In doing this, this forum covers a broad spectrum of contributions from the conception of technical solutions to the assessment of the benefits in particular populations. The aim is for users of these smart environments to be unaware of the underlying technology, while reaping the benefits of the services it provides. Devices embedded within the environment are aware of the people's presence and subsequently react to their behaviors, gestures, actions and context.

During the last years, the interest in ubiquitous computing and ambient intelligence has grown considerably, due to new challenges posed by society, demanding highly innovative services for several application domains, such as vehicular ad hoc networks, ambient-assisted living, e-health, remote sensing, home automation and personal security. The COVID-19 pandemic has made us not only more aware of the ubiquity of information technologies in our daily life but also of their need to seamlessly support our everyday activities.

We are concerned with the sound development of UC and AmI as the only way to properly satisfy the expectations around this exciting intersection of information, communications and control technologies. Therefore, this UCAmI edition involves research work in five tracks: AmI for health and A3L, Internet of everything and sensors, smart environments, human–computer interaction and data science.

We received 107 submissions for this 15th edition of UCAmI authored by 207 researchers from 18 countries. A total of 297 reviews were performed, reaching the high average of 2.19 reviews per submission. We would like to thank all the authors who submitted their work for consideration, as well as the reviewers who provided their detailed and constructive reviews. Many thanks also to the track chairs for the great commitment shown in organization and execution of the papers reviewing process.

Finally, we are happy to return to the same place of the 2011 edition of the conference: Riviera Maya (Mexico).

November 2023

José Bravo
Gabriel Urzáiz

Organization

Our Staff

José Bravo (General Chair)	University of Castilla-La Mancha, Spain
Gabriel Urzaiz (Local Chair)	Anáhuac Mayab University, Mexico

Steering Committee

José Bravo, Spain
Pino Caballero, Spain
Macarena Espinilla, Spain
Jesús Favela, Mexico
Diego López-De-Ipiña, Spain
Chris Nugent, UK
Sergio F. Ochoa, Chile
Ramón Hervás, Spain
Gabriel Urzaiz, Mexico
Vladimir Villareal, Panama
Jesús Fontecha, Spain
Iván González, Spain

Organization Committee

Cosmin Dobrescu, Spain
David Carneros, Spain
Laura Villa, Spain
Luis Cabañero, Spain
Tania Mondéjar, Spain
Esperanza Johnson, Spain
Alejandro Pérez, Spain
Brigitte Nielsen, Panama
Paloma Bravo, Spain

Track Chairs

AmI for Health & (A3L) (Ambient, Active & Assisted Living)

Jesús Fontecha, Spain
Ian Cleland, UK

Smart Environment

Macarena Espinilla, Spain
Kåre Synnes, Sweden
Chris Nugent, UK

Internet of Everything (IoT + People + Processes) and Sensors

Joaquín Ballesteros, Spain
Cristina Santos, Portugal

Data Science

Marcela Rodríguez, Mexico
Alberto Morá, Mexico

Human–Computer Interaction

Gustavo López, Costa Rica
Sruti Subramanian, Norway

Satellite Events

International Workshop on Energy Aware Systems, Communications and Security

Mauro Migliardi, Italy
Francesco Palmieri, Italy

Program Committee

Adrian Lara	UCR
Adrian Sánchez-Miguel Ortega	Universidad de Castilla-La Mancha
Alberto Morán	UABC

Alejandro Pérez Vereda	Universidad de Castilla-La Mancha
Alessio Merlo	University of Genova
Alireza Souri	Haliç University
Allan Berrocal	Universidad de Costa Rica
Andres Diaz Toro	UNAD
Andrés Oliva	Anáhuac Mayab University
Antonio Robles-Gomez	UNED
Antonio Albín Rodríguez	Universidad de Jaén
Arcangelo Castiglione	University of Salerno
Arfat Ahmad Khan	Department of Computer Science, College of Computing, Khon Kaen University
Beatriz Garcia-Martinez	Universidad de Castilla-La Mancha
Borja Bordel	Universidad Politécnica de Madrid
Bruno Carpentieri	University of Salerno
Carlo Ferrari	University of Padova
Carlos Rovetto	Universidad Tecnológica de Panamá
Carlos Aguilar Avelar	UABC
Carlos E. Galván	Universidad Autónoma de Zacatecas
Carmelo Militello	Italian National Research Council (CNR)
Carmen Martinez Cruz	University of Jaen
Chris Nugent	Ulster University
Colin Shewell	Ulster University
Constantin Cosmin Dobrescu	Universidad de Castilla-La Mancha
Cristiana Pinheiro	University of Minho
Cristina Santos	University of Minho
Cristina Ramirez-Fernandez	TecNM/I.T. de Ensenada
David Carneros-Prado	Universidad de Castilla-La Mancha
David Gil	University of Alicante
Davide Zuccarello	Politecnico di Milano
Dionicio Neira Rodado	Universidad de la Costa
Eduardo Barbará	Anáhuac Mayab University
Elena Navarro	Universidad de Castilla-La Mancha
Ernesto Lozano	CICESE
Ernesto Vera	UABC
Esperanza Johnson	Høgskolen i Innlandet
Fabio Lopes	Universidade Presbiteriana Mackenzie
Fabio Salice	Politecnico di Milano
Federico Cruciani	Ulster University
Federico Botella	UMH
Francesco Palmieri	University of Salerno
Francisco Flórez-Revuelta	University of Alicante
Francisco Javier Cabrerizo	University of Granada

Luis Cabañero	University of Castilla-La Mancha
Luis Quesada	Universidad de Costa Rica
Luis Pellegrin	UABC
Luís Moreira	University of Minho
Luis A. Castro	Instituto Tecnologico de Sonora (ITSON)
Macarena Espinilla	University of Jaén
Manuel Fernández Carmona	Universidad de Málaga
Marcela Rodriguez	UABC
María Martínez Pérez	University of A Coruña
Maria del Pilar Angeles	IIMAS, UNAM
Matias Garcia-Constantino	Ulster University
Matteo Venturelli	Politecnico di Milano
Matthew Burns	Ulster University
Mauro Migliardi	Universita' degli Studi di Padova
Mauro Iacono	Campania University L. Vanvitelli
Mercedes Amor	Universidad de Málaga
Michele Mastroianni	University of Salerno
Muhammad Asif Razzaq	Fatima Jinnah Women University
Nuno Ferrete Ribeiro	University of Minho
Oihane Gómez-Carmona	University of Deusto
Rafael Pastor-Vargas	UNED
Ramón Hervás	University of Castilla-La Mancha
Rene Navarro	University of Sonora
Rosa Arriaga	Georgia Tech
Rubén Domínguez	Anáhuac Mayab University
Saguna Saguna	Luleå University of Technology
Salvatore Source	Università degli Studi di Enna "Kore"
Sandra Nava-Munoz	UASLP
Sara Comai	Politecnico di Milano
Sara Cerqueira	University of Minho
Sergio Ochoa	University of Chile
Solomon Sunday Oyelere	Luleå University of Technology
Sruti Subramanian	Norwegian University of Science and Technology
Tania Mondéjar	University of Castilla-La Mancha
Tatsuo Nakajima	Waseda University
Unai Sainz Lugarezaresti	University of Deusto
Vaidy Sunderam	Emory University
Vincenzo Conti	University of Enna KORE
Vladimir Villarreal	Universidad Tecnológica de Panamá
Wing W. Y. Ng	South China University of Technology
Yarisol Castillo	Universidad Tecnológica de Panamá
Zaheer Khan	University of the West of England, Bristol

Contents

About the Editors

Dr. José Bravo is Full Professor in Computer Science in the Department of Technologies and Information Systems at Castilla-La Mancha University, Spain, and Head of the Modelling Ambient Intelligence Research Group (MAmI, mami- lab.eu). He is involved in several research areas such as ubiquitous computing, ambient intelligence, ambient assisted living, context-awareness, Internet of things, mobile computing and m-Health. He is an author of over 37 JCR articles and the main researcher on several projects. H-Index (Scopus). - 22, H-Index (Google Scholar). – 30. Dr. Bravo supervised 7 PhD and over 45 Computer Science undergraduate theses. Since 2003, José Bravo has been the organizer of the International Conference on Ubiquitous Computing & Ambient Intelligence (UCAmI).

Dr. Gabriel Urzáiz received his BS in Computer Engineering at the National Autonomous University of Mexico and his PhD in Advanced Computer Technologies at the Castilla-La Mancha University. His research activity is focused on computer networks, mainly for the integration of heterogeneous networks and their application in ambient intelligence and ubiquitous computing. He is the author of several conference and journal papers, and he has also participated as a reviewer and guest editor. He has been the director of the Computer Science School of the Anahuac Mayab University in Mexico, a research professor and a postgraduate academic coordinator. His current position is as a full-time professor in the Engineering and Exact Sciences Division, primarily focused on teaching and student mentoring.

Internet of Everything (IoT + People + Processes) and Sensors

Distributed Crowdsensing Based on Mobile Personal Data Stores

Alejandro Perez-Vereda[1(✉)], Luis Cabañero[1], Nathalie Moreno[2], Ramon Hervas[1], and Carlos Canal[2]

[1] Universidad de Castilla - La Mancha, Ciudad Real, Spain
{alejandro.pvereda,luis.cabanero,ramon.hlucas}@uclm.es
[2] ITIS Software, Universidad de Málaga, Malaga, Spain
{nmv,carloscanal}@uma.es

Abstract. Users' personal information is one of the most important actives for nowadays enterprises. Knowing user preferences allows to offer personalized interactions and obtain more high-value information. In this context, crowdsensing shows as a technique that aims to collect information about the users and their Internet of Everything (IoE) environment. Personal smartphones are the devices that act as the interface between people and the IoE. However, most of the related works in the literature consider smartphones as mere sensors that gather user data, which is then transferred to the crowdsensing requester. As an alternative, in this paper we propose a distributed crowdsensing platform based on a extension of the Digital Avatars framework for Mobile Collaborative Social Computing. In our proposal, smartphones are responsible for compiling and keeping the digital avatar or virtual profile of each of the users participating in the crowdsensing activity. Based on these avatars, the framework is extended to provide a distributed platform for both the dissemination and the aggregation of the results of the activity, granting users with privacy and ownership of their personal data. Our proposal also takes into account trust and user reputation by means of subjective logic. The proposed system is tested and validated through a proof of concept.

Keywords: Crowdsensing · Distributed Crowdsensing · Virtual Profiles · Digital Avatars · Personal Data Store

1 Introduction

Nowadays, companies from the social networks realm obtain a vast amount of data from their users. These data are employed to improve the quality of the services provided, adapting them to user's needs and preferences, and also for

This work has been funded under the Spanish research projects RTI2018-098780-B-I00, which funds the PRE2019-089614 predoctoral contract, PID2021-125527NB-I00, and TED2021-130523B-I00.

J. Bravo and G. Urzáiz (Eds.): UCAmI 2023, LNNS 841, pp. 3–15, 2023.
https://doi.org/10.1007/978-3-031-48590-9_1

analytics and marketing purposes, among others. This evidences the importance of crowdsensing in current IT societies. Crowdsensing is an emerging paradigm for information gathering, based on the availability of people's mobile devices to collect and transmit information about any phenomena of interest [5].

However, most IT systems are based on centralized architectures [11], where personal information is transferred to the company running the system. Mobile devices, and therefore their users, are conceived as passive entities that collect and transmit information to centralized servers where it is processed. This creates a gap between all this information and its fair owners —the users—, who cannot decide how their data is managed, neither who has granted access to it.

In previous works, some of the authors of this paper presented the Digital Avatars framework [14]. Digital Avatars leverages the computing, storage, and communication capabilities of current smartphones, together with their pervasive presence, for making them personalized interfaces to their owners, offering services to third parties based on the information contained in a *digital avatar* or virtual profile of the user stored in the smartphone. The avatar is inferred by the smartphone from the activities of the user, who keeps the control to decide how and with whom sharing their data.

In this paper, we present an extension of the Digital Avatars framework to support crowdsensing activities in a distributed manner. In particular, we build a loosely coupled crowdsensing platform in which users and their smartphones become the main players, disseminating the tasks among the participants and aggregating the results, building collective profiles from the digital avatars of each of the participants. Additionally, a trust management system is integrated in the platform to address privacy and security concerns. A proof of concept has been implemented to test the platform and validate the proposal in terms of performance and efficiency. The source code of the proposal, together with all the validation data is available in a public repository[1].

The structure of this paper continues with the presentation in Sect. 2 of the case study used a as proof of concept of our proposal. In Sect. 3, we revise the state of the art and provide related works. Then, Sect. 4 briefly describes the Digital Avatars framework. Section 5 deals with our approach to trust and uncertainty. The validation of the platform is available at Sect. 6. Finally, Sect. 7 draws the conclusions and discusses future works.

2 Motivating Scenario

Anna and John are two vegetarian food enthusiasts who would like to open a healthy food restaurant in a well-known neighborhood in the city of Malaga. The area is mainly inhabited by a young population with average purchasing power. Given the volume of fast food restaurants in the area, one might think that the eating habits of its inhabitants are not too healthy. Therefore, before venturing into this important economic investment, Anna and John have decided to carry

[1] Public repository of the proposal: https://github.com/apvereda/UCAmI2023-CrowdSensing.git

out a market study. Their goal is to learn the profile and lifestyle habits of their potential customers. This market study would include questions that help them to identify the age range of their potential clients, the frequency with which they go to this type of establishments, the amount of money they usually spend, what dishes they would like to see included in the menu, if they consider that a good wine list is something relevant, if they would be interested in an online ordering service, etc.

Most of the services that conduct surveys like the one required in this scenario do not guarantee the total privacy/anonymity of the users, nor do they allow any control over the network of people to whom the survey is delivered. Anna and John know that word of mouth is a very effective marketing strategy that helps build a loyal fan base. That is why their initial objective is to send the survey to their direct contacts, those they trust the most, or have the best reputation in the gastronomic field, and let them be the ones to spread the survey following the same strategy. In this way, the initial reluctance of many people to collaborate with this type of survey could be overcome and a higher rate of participation would be achieved.

In the next section, we describe the alternatives found in the literature for this and other similar crowdsensing problems, focusing on aspects that need to be improved and that we address in our proposal.

3 Related Work

Prior to the introduction of our crowdsensing platform, it is relevant to explore the state of art to frame it properly. In [7], many crowdsensing systems are explored to establish different perspectives to classify them. One of these perspectives is defined by how the user is involved in data retrieving, which establishes two main alternatives: participative, in which the user actively decides which data is going to be shared, and opportunistic, in which the data is gathered in a transparent way. There are also intermediate positions, which is where we aim our platform to be: following a participative approach, but allowing automatic response in some cases. Another taxonomy made by the aforementioned paper classifies the systems depending on the scale, and propose three types: group, community and urban. Our aim is to focus on communities, that will be reached by contacts networks, and including constrains for participation, so their spread will be limited. The last relevant aspect to pinpoint of the proposed system is its distributed nature, which is not something very common in crowdsensing.

Crowdsensing can be applied to many purposes, such as groundwater contamination detection [17], public transport flow optimization [15], or mapping the noise levels on a city [3]. As it can be seen, there is high variability in the uses of crowdsensing and it is important to decide the scope of the system before building it. Our proposal consists of a participative and distributed crowdsensing platform oriented towards community-scale surveys and other similar activities, where the dissemination of the tasks relies on community and personal networks and users' reputation, as it will be shown. However, distributed crowdsensing has not been explored in depth, although there are some works worth mentioning. The first of them is [10], in which a edge-computing approach using intermediate nodes between the sensors and the main server is presented. This article shows how using distributed architectures is beneficial for crowdsensing, but the way the work is distributed is not between smartphones. Another relatable work is [16], which uses an opportunistic approach to combine data from multiple devices whenever they are close enough, keeping anonymity by using cryptographic techniques. Despite both works have a distributed orientation, these approaches differ from ours in the use a central server that compiles all the information gathered.

4 Digital Avatars for Distributed Crowdsensing

The Digital Avatars framework is built over the concept of virtual profile or *digital avatar* of a user to offer services based on it. The avatar can be understood as a personal data store, which is inferred and kept uniquely in the user's smartphone, and contains personal information about the user, their contacts, habits, and any other purpose-specific information. The data store is represented by JSON documents kept in a NoSQL database installed in the smartphone.

The framework provides tools and services for third parties to access the avatar and to interact with the user, and it is developed as a single Android app that runs in the background in a seamless way. Thus, the digital avatar becomes a proxy of the user with the Internet of Everything (IoE) environment, allowing them to keep control over their information through tailored privacy settings and rules for determining the avatar behaviour and inference.

Interaction with the avatar is achieved by means of two specific components: The Inference Engine, which is based on Complex Event Processing (CEP) technology [9] for specifying the rules that determine how information is inferred and stored in the avatar, and the Execution Core, which employs a Beanshell[2] Java interpreter which grants the smartphone the capability of runtime execution of purpose-specific code.

For more information on the Digital Avatars framework, the reader may refer to [13,14]. In the rest of this section, we describe how we have extended the Digital Avatars framework for building the distributed crowdsensing platform which is the main contribution of this paper. The aim of this extension is using the framework to provide a new service for third parties to perform surveys, polls, or analytics on some user information of interest. Due to the nature of the framework, this functionality is achieved in a loosely coupled and distributed manner, being the smartphones in charge of disseminating the surveys and aggregating their results, protecting users' privacy.

[2] BeanShell: https://github.com/beanshell/beanshell

4.1 Distributed Crowdsensing

Crowdsensing consists in acquiring some information from a considerable number of users in order to obtain some aggregated result or analytic. A crowdsensing task may adopt the form of a survey for acquiring some specific opinion or data from the user. When a crowdsensing task notification arrives, it is added to the list of uncompleted tasks in the platform, so the user can consult all active crowdsensing tasks and decide whether to accept and answer them. A task may require the user to answer a survey, or it just fetch some user information already stored in the digital avatar, not requiring the user intervention for more than trusting the task sender.

In our proposal the module in charge of controlling the flow of the crowdsensing tasks is the Inference Engine. As mentioned above, CEP is a technology that allows to analyze and correlate in real time raw data coming from multiple sources, identifying patterns to discover complex events or detect situations of interest. For running CEP in smartphones, we employ Siddhi [18], which is an open source CEP engine successfully extended to Android devices by the use of plugins. The behavior of the engine relies on the applications running on it, each one consisting in a set of CEP rules for detecting domain-specific event patterns and react to these situations depending on incoming data. There can be several applications running at the same time, using different sources of data and with different purposes. The Siddhi IO Android plugin endows the engine with capabilities inherent to smartphones, like the management of built-in sensors by the use of *sources*, which are simple event or data producers, and *sinks*, which corresponds to actuators.

The Inference Engine is in charge of receiving crowdsensing tasks, and also of disseminating them over the contacts of the user, from now on called *worker*. An example of how a crowdsensing task spread over the community of workers can be seen in Fig. 1. Each task defines the number of level (i.e. the depth of the crowdsensing tree). This number is decremented by 1 with each dissemination over a worker's contacts. This way, when a worker receives a task whose depth level is 0, it means that this worker is a leaf of the crowdsensing tree, so the task is not disseminated anymore, and the only thing the worker has to do is answer the task and send back the results. When a worker receives a task with depth level bigger than 0, the worker can react in two ways:

- The worker joins the task and marks it as in progress. Then, the worker spreads the task to their own contacts, decreasing its level by 1. When the worker receives back the results from their contacts, it notifies the aggregated result with its own answer (in the white notes in the figure) to the worker from whom it received the task.
- The worker declines or ignores the task. These are represented in the figure by a red cross over the arrow. The reason for that may be that the worker is not interested in the task or it does not trust its origin, but also that this same task has been already received from another contact of the worker.

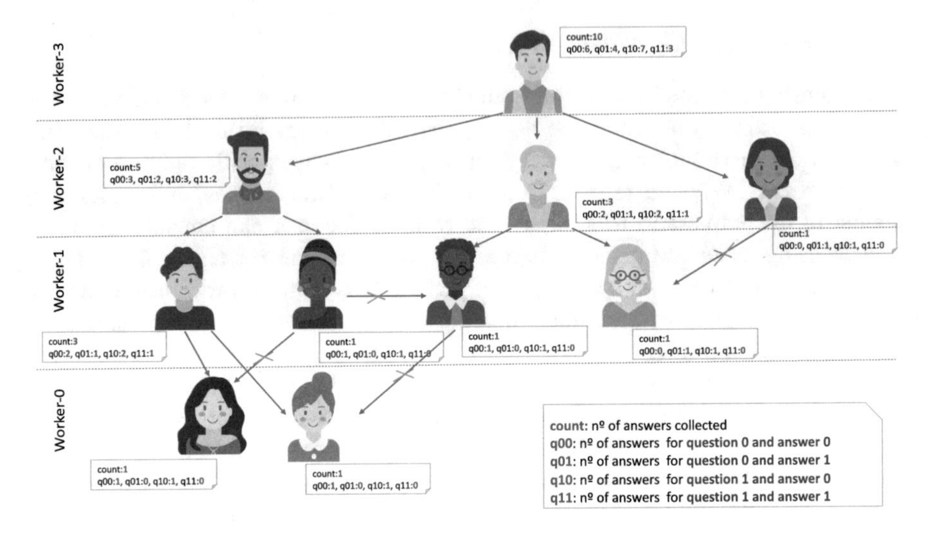

Fig. 1. Example crowdsensing task —consisting of two yes/no questions— over a contact network. The graph shows how the task is iteratively disseminated over the workers' contacts and the aggregation of the results.

This behavior of the Inference Engine is specified by a Siddhi application —a set of CEP rules— which receives crowdsensing tasks as incoming events and disseminate them. We use push notifications to spread the tasks over the smartphones, interconnecting the corresponding Siddhi engines. In particular, we use the OneSignal[3] platform for sending push notifications. OneSignal is a system for mobile push, web push, email, and in-app messages based on Google's Firebase Messaging service. It provides a cloud service with an Android native API and offers a free tier with enough capacity for testing our implementation. To integrate the push notifications service, we have extended Siddhi with a listener from the OneSignal API —implemented as a new Siddhi *source*— in order to listen to incoming push notifications. This *source* handles the notifications and extracts all the information they contain. In a similar way, the set of *sinks* must be extended for sending push notifications to the contacts of the workers.

We can see how a crowdsensing task notification looks like in the example in Code 1. In the *additionalData* field of the notification we find the task attributes; *appID* is a namespace used internally to refer to the Siddhi application managing the crowdsensing tasks (as several Siddhi applications with different purposes may be running together in the engine), *tokenID* is used to verify the identity of the task sender, *role* specifies the depth level of the task, *pollID* is the task identifier, while the *survey* field consists in a JSON survey to be answered by the user in case of requiring more data than that actually stored in the avatar, *callback* indicates to whom send the aggregated results of the task, which can be either another worker OneSignal ID, or an URL to which upload the result

[3] OneSignal: https://onesignal.com/.

for the first level of workers who received the task directly from the crowdsensing requester. Finally, *timeout* and *script* have to do with the Execution Core component, that also needs to be integrated in the flow of the crowdsensing framework, as we explain below.

```
1  {...''notification": {
2        ''payload": {
3            ''notificationID": ''d5hs6278-n884-...-5b3hd727c2a"
   ,
4            ''title": ''example@email.es",
5            ''body": ''",
6            ''additionalData": {
7                ''appID": ''DA-PollReceiver",
8                ''tokenID": ''some.Firebase.JWT.Token",
9                ''role": ''worker-2",
10               ''timeout": ''1800000",
11               ''pollID": ''arbitraryPolln1",
12               ''script": ''bit.ly/xxx",
13               ''survey":
14                   ''[{''questionText":''Like veg rest?",
15                   ''answer1":''Yes",''answer2":''No"},
16                   {''questionText":''Have you tried?",
17                   ''answer1":''Yes",''answer2":''No"}]",
18               ''callback": ''c53a5126-d987-...-09a9f5ce5c70"
19           },
20           ...
21       }}...}
22
```

Code 1. Example of push notification for a crowdsensing task.

The Execution Core is responsible for fetching and aggregating the information in our distributed crowdsensing platform. The Execution Core is able to run Beanshell Java scripts, implementing the required behaviour for a given purpose of use of the Digital Avatars framework. In our case, the script to execute is provided in the crowdsensing task notification. A task script contains instructions to access the data and the results for the crowdsensing task in the digital avatar of the worker, and to aggregate the results of the received answers from the worker's contacts.

Indeed, when a worker receives a crowdsensing task, it both completes the corresponding survey, and spreads the task through their contacts. The *timeout* attribute of the task states the time to wait after the task is disseminated until the worker executes the script to aggregate the results received. In each dissemination step, the timeout is reduced 1/level times so that the worker has time to answer the survey, aggregate the results, and send back the response to the worker from which it received the task. Answers arriving after the timeout has passed will be ignored.

All the source code of the distributed crowdsensing platform developed over the Digital Avatars framework presented in this paper can be found at GitHub[4]. This implementation has been used for the validation presented in Sect. 6. Further considerations on security and privacy on the dissemination of the crowdsensing tasks are addressed in the following section.

5 Trust in Crowdsensing Tasks

Trust refers to the degree of reliability (trustworthiness) we assign to people to perform an action (*functional* trust) [2]. Such a relationship does not need to be symmetric. Trust is also context-dependent, which means that someone does not need to be trusted in all situations. Hence, it is not absolute and must be specified within a *scope* [6]. Finally, trust is *subjective*, and it is normally conditioned by uncertain factors [8]. In this section we explain how we represent trust and subjective opinions in our crowdsensing approach.

5.1 Subjective Logic

Degrees of trust or confidence have been traditionally modeled using probabilities, and reasoning about trust has been accomplished using probability theory [4]. However, this approach shows evident limitations when it comes to representing subjective opinions for which users cannot easily express their ignorance about the facts they are considering, or their inability to assign an accurate probability to a fact. Forcing users to set probabilities to express their opinions could lead to unreliable conclusions [12].

Subjective Logic [8], is an extension of probabilistic logic that explicitly takes uncertainty into account. Subjective opinions express beliefs about the truth of propositions under degrees of uncertainty. They can also indicate confidence, or trust, on a given statement and this is what makes them suitable in our context.

Let x be a Boolean predicate. A binomial *opinion* about the truth of x is defined as a quadruple (b_x, d_x, u_x, a_x) where b_x (*belief*) is the degree of belief that x is true, d_x (*disbelief*) is the degree of belief that x is false, u_x (*uncertainty*) is the degree of uncertainty about x, and a_x (*base rate*) is the prior probability of x. These values satisfy that $b_x + d_x + u_x = 1$, and $b_x, d_x, u_x, a_x \in [0, 1]$.

Intuitively, the *base rate* of an opinion represents the *objective* probability that can be assigned to the statement using *a priori* evidences or statistical estimates, whilst the other elements of the tuple represent the *subjective* degrees of belief, disbelief, and uncertainty about the statement. Thus, regardless of the value of the prior probability, different belief agents can express subjective opinions about the statement, including their degree of uncertainty.

To represent and operate with Subjective Logic values, in [12] we defined the datatype SBoolean that extends booleans with uncertainty information, and

[4] Public repository of the proposal: https://github.com/apvereda/UCAmI2023-CrowdSensing.git.

provides all corresponding operations. The embedding of a probability c representing a confidence into type SBoolean is achieved by assigning the opinion $(c, 1-c, 0, c)$ to x. Considering the embedding of type Boolean into Probabilities, we have that Boolean values true and false correspond, respectively, to opinions $(1, 0, 0, 1)$ and $(0, 1, 0, 0)$. Examples of the use and application of Subjective logic in models represented with UML/OCL can be found in [12].

5.2 Considering Trust in Crowdsensing

In our context, trust refers to the degree of confidence we have in a worker. When a worker receives a crowdsensing task she disseminates it through her contacts. For that it is necessary to establish a filter so that the task is only sent to those contacts that the worker trusts the most. When a user A adds a new contact B, our crowdsensing platform asks A about her trust in B. The rating is based on a star rating system and it translated into a representation in subjective logic and stored in her digital avatar. It could happen that A has had no previous experience with B so her functional confidence is initialized to an opinion where uncertainty is maximal and the degree of belief and disbelief are 0 (hence the importance of uncertainty). Then, if A interacts with B in a positive way (B responds to the poll requests made by A), this functional trust is modified, reducing uncertainty and increasing the degree of belief. The platform uses this functional trust as follows:

- Workers disseminate task proposals only to those contacts whose trust exceeds a certain threshold.
- Workers only answer and send back the result of task proposals that come from workers they trust.
- While using the platform, a worker updates these values based on the last interactions they have had other users.

With this mechanism we try to maximize the degree of response and interaction of users in this type of surveys that are generally not attended due to the distrust they generate. People are more likely to trust and respond to requests that come from other people they trust.

6 Validation

In this section we present the validation carried out on our crowdsensing platform. First, as the system runs in background in the smartphones, we have conducted several experiments for evaluating the efficiency and performance of the proposal. Second, we have installed the platform in ten personal smartphones, and checked its functionality performing crowdsensing tasks.

In a previous work of some of the authors of this paper [1], we showed that there are many situations in which a smartphone-based architecture performs better and more efficiently for building Collaborative Social Computing applications, as those of crowdsensing are. Nevertheless, it is important to evaluate also how specific elements of the crowdsensing framework, such as push notifications, and running a CEP engine affects resource consumption in the smartphones. To this extent, we here describe the results of carrying performance tests on the OneSignal based communications used in the proposal, and an overall evaluation of energy consumption for the crowdsensing framework. For the tests, we used the tools Android Profiler[5], which provides real-time data about how an app uses CPU, memory, network, and battery resources, and Battery Historian[6], which visualizes battery related information of either the whole Android system or any given application. All the results obtained and plots generated can be found at the evaluation folder in the GitHub companion repository of this paper[7]

For evaluating the robustness of the communication architecture under stress situations we measured the time required for transmitting a number of push notifications between Android smartphones, and whether some of these notifications were lost in the way. We also measured their impact on CPU and battery usage. We repeated these experiments ten times, sending out at once packets of 100, 300 and 500 messages assuming good network conditions. The results obtained in these tests show that the messages take a similar time to deliver, being the mean time between messages of 0.126 ± 0.029 seconds through all the tests made. Additionally, if we take a look at energy consumption does not surpass the *light* level marked by the analysis tool, while the CPU usage peaks does not even reach a 20% of the computing capabilities of the smartphone. One thing to consider is that when performing the 500 messages tests, we started to experiment some losses of no more than 10 messages, suggesting that the system reached its limit.

Furthermore, battery consumption is a critical issue in mobile devices and one of the concerns that may prevent users to install a new app in their phones. Thus, we have also performed tests for estimating battery usage running the crowdsensing framework for 3 and 6 h. During these executions push notifications were received by the application at a rate of about one per minute, and the OneSignal message handler task was activated 307 times for receiving messages during the run. This produced a usage of a 0.05% of battery resources, which is a very low figure. Hence, we conclude that even in stress situations, there is no evidence of notable impact on smartphone resources.

For the experiment with real users, we implemented the crowdsensing task about a vegetarian restaurant described in Sect. 2. The network of contacts is that depicted in Fig. 1, involving ten different workers with their own personal smartphones, with Android versions ranging from 8 to 13. The level of the survey was set to 4. This implies that some workers receive the same task from several contacts. The survey was simplified to two yes/no questions which the workers were asked to answer. On a first stage, the task is sent to just one worker —that in the top of the figure—, who spread it over his contacts, decreasing the timeout

[5] Android Profiler: https://developer.android.com/studio/profile/android-profiler?hl=es-419

[6] Battery Historian: https://github.com/google/battery-historian

[7] Proposal repository: https://github.com/apvereda/UCAmI2023-CrowdSensing.git

and level of the task at each step. Then, when the corresponding timeout reaches, if the worker has accepted the task, she aggregates her results with the answers received from her contacts, and sends the results back to the higher-level worker, up to the top of the tree.

We carried out the experiment three times, trying to represent real world situations to show how the system worked. For that, we caused some smartphones to fail on the spread of the task due to temporary connection loss, or simply to decide not to answer the survey. In all the experiments, the crowdsensing platform performed as expected, without uncontrolled errors and returning the aggregated result of the participating users. No message was lost during the experiments and the communications were fast. This demonstrates the soundness of the system, considering that ours is a loosely-coupled distributed approach, and that reaching the full network and obtaining all the answers is not granted, an inherent characteristic of our approach. After the experiments, we collected and combined the execution traces of all the smartphones involved, which show the actions taken by each worker and the exchange of messages between the smartphones. These traces can be also found in the companion repository.

7 Conclusions and Future Work

In this work, we have presented an innovative crowdsensing platform, oriented towards community-scale crowdsensing activities, where the dissemination of the tasks relies on personal networks. The proposal follows a loosely coupled approach which puts the users in the center of the picture, giving them the authority to manage their own generated information through a personal virtual profile allocated in their smartphone. In contrast with related works, our proposal is a completely distributed solution that establishes direct communication between workers' devices, without requiring a central server. This reinforces anonymity, privacy, and security as all the data are aggregated by the workers in their own smartphones. The requester of the crowdsensing task only receives the computed results for the survey and the number of users involved. Apart from using it for interactive surveys, the platform is able to perform polls without user intervention, as the information enquired may be already stored in the avatar.

Furthermore, we have presented a trust management system for this crowdsensing platform, based on subjective logic. This trust system helps encouraging and engaging the users to join the crowdsensing tasks, as they will receive them from a contact they trust in.

As future work, we plan to extend trust management with reputation considerations, giving support to decision-making for accepting incoming tasks. Additionally, a more complete validation of the system is needed, implying a larger network of workers. For this purpose, we will employ tools for smartphone virtualization, making easier the validation of the system.

References

1. Berrocal, J., et al.: Early evaluation of mobile applications' resource consumption and operating costs. IEEE Access **8**, 146648–146665 (2020). https://doi.org/10.1109/ACCESS.2020.3015082
2. Braga, D.D.S., Niemann, M., Hellingrath, B., Neto, F.B.D.L.: Survey on computational trust and reputation models. ACM Comput. Surv. **51**(5), 1–40 (2018). https://doi.org/10.1145/3236008
3. Buwaya, J., Rolim, J.: NoiseBay: a real-world study on transparent data collection. In: Proceedings of the 14th International Conference on Contemporary Computing, pp. 493–501 (2022). https://doi.org/10.1145/3549206.3549325
4. de Finetti, B.: Theory of Probability: A Critical Introductory Treatment. Wiley, Hoboken (2017)
5. Ganti, R.K., Ye, F., Lei, H.: Mobile crowdsensing: current state and future challenges. IEEE Commun. Mag. **49**(11), 32–39 (2011). https://doi.org/10.1109/MCOM.2011.6069707
6. Grandison, T., Sloman, M.: A survey of trust in internet applications. IEEE Commun. Surv. Tutorials **3**(4), 2–16 (2000). https://doi.org/10.1109/COMST.2000.5340804
7. Guo, B., et al.: Mobile crowd sensing and computing: the review of an emerging human-powered sensing paradigm. ACM Comput. Surv. **48**, 1–31 (2015). https://doi.org/10.1145/2794400
8. Jøsang, A.: Subjective Logic. Artificial Intelligence: Foundations, Theory, and Algorithms. Springer, Cham (2016). https://doi.org/10.1007/978-3-319-42337-1
9. Luckham, D.C.: Event Processing for Business: Organizing the Real-Time Enterprise. Wiley, Hoboken (2011)
10. Marjanović, M., Antonić, A., Žarko, I.P.: Edge computing architecture for mobile crowdsensing. IEEE Access **6**, 10662–10674 (2018). https://doi.org/10.1109/ACCESS.2018.2799707
11. Minerva, R., Crespi, N.: Unleashing the disruptive potential of user-controlled identity management. In: 2011 Technical Symposium at ITU Telecom World (ITU WT), pp. 1–6 (2011)
12. Muñoz, P., Burgueño, L., Ortiz, V., Vallecillo, A.: Extending OCL with subjective logic. J. Object Technol. **19**(3), 3–1 (2020). https://doi.org/10.5381/jot.2020.19.3.a1
13. Pérez-Vereda, A., Canal, C., Pimentel, E.: Modelling digital avatars: a tuple space approach. Sci. Comput. Program. **203**, 102583 (2021). https://doi.org/10.1016/j.scico.2020.102583
14. Perez-Vereda, A., Hervas, R., Canal, C.: Digital avatars: a programming framework for personalized human interactions through virtual profiles. Pervasive Mob. Comput. **87**, 101718 (2022). https://doi.org/10.1016/j.pmcj.2022.101718
15. Plašilová, A., Procházka, J.: Crowdsensing technologies for optimizing passenger flows in public transport. In: 1st International Conference on Advanced Innovations in Smart Cities (ICAISC) (2023). https://doi.org/10.1109/ICAISC56366.2023.10085515
16. Reinhardt, D., Manyugin, I.: OP4: an OPPortunistic privacy-preserving scheme for crowdsensing Applications. In: IEEE 41st Conference on Local Computer Networks (LCN), pp. 460–468 (2016). https://doi.org/10.1109/LCN.2016.75

17. Shang, L., et al.: CrowdWaterSens: an uncertainty-aware crowdsensing approach to groundwater contamination estimation. Pervasive Mob. Comput. **92**, 101788 (2023). https://doi.org/10.1016/j.pmcj.2023.101788
18. Suhothayan, S., Gajasinghe, K., Loku Narangoda, I., Chaturanga, S., Perera, S., Nanayakkara, V.: Siddhi: a second look at complex event processing architectures. In: ACM Workshop on Gateway Computing Environments, pp. 43–50. ACM (2011). https://doi.org/10.1145/2110486.2110493

FreeDSM: An Open IoT Platform for Ambient Light Pollution Monitoring

Daniel Boubeta[1(✉)], Carlos Dafonte[1], Eduard Masana[5], Ana Ulla[3,4], Alejandro Mosteiro[1], and Minia Manteiga[2]

[1] CIGUS CITIC - Department of Computer Science and IT, University of A Coruña (UDC), Campus de Elviña s/n, 15071 A Coruña, Spain
daniel.boubeta@udc.es

[2] Institut d'Estudis Espacials de Catalunya (IEEC), c. Gran Capitá, 2-4, 08034 Barcelona, Spain

[3] Applied Physics Department, Universidade de Vigo (UVIGO), Campus Lagoas-Marcosende, s/n, 36310 Vigo, Spain

[4] IFCAE - Instituto de Física e Ciencias Aeroespaciais, Universidade de Vigo, Campus de As Lagoas, 32004 Ourense, Spain

[5] CIGUS CITIC - Department of Nautical Sciences and Marine Engineering, University of A Coruña (UDC), Paseo de Ronda 51, 15011 A Coruña, Spain

Abstract. Light pollution is one of the fastest growing environmental problems in recent years, mainly affecting urbanized areas worldwide. It goes beyond the difficulty of the astronomical observation, causing a deep impact on the balance of ecosystems, wildlife and human health. To deal with this complex situation, it is necessary to increase public awareness and to demand from all implicated stakeholders efficient technical solutions to stop and mitigate unwanted light pollution effects. A continous monitoring of light pollution levels is also mandatory. In this context, we propose a new device called FreeDSM, an IoT based photometer to measure the quality of the dark sky at night. FreeDSM will be easy-to-use and easy-to-build, so every interested citizen can create their own and collaborate with the project. The main component of the photometer is an ESP32C3 microcontroller, a cheap, small, low consumption and powerful chip. Also, a TSL2591 optical sensor, of high sensitivity and large dynamic range, is included. Thanks to the proposed design, additional positioning or ambiance sensors could be incorporated. To handle all these components, the ESP32 firmware is implemented through Tasmota, an open-source home assistant solution for most of commercial sensors. And all the information to be gathered by the FreeDSM devices will be made public via a platform built using FIWARE, an open framework for IoT solutions. By doing so, this solution aims to standarization of IoT collected data, easing users own new applications development and public awareness increase of the problem. Our proposed approach is expected to empower individuals to take action against unnecessary outdoor lightning.

Keywords: Light Pollution · Open Hardware · Citizen Science

J. Bravo and G. Urzáiz (Eds.): UCAmI 2023, LNNS 841, pp. 16–24, 2023.
https://doi.org/10.1007/978-3-031-48590-9_2

1 State of the Art

1.1 Light Pollution Problem

Light pollution is understood as the alteration of the natural brightness of the night sky due to the excess or improper use of artificial light sources at unnecessary times, directions or intensities. It is one of the environmental problems that grows the most each year [11], and yet, it is one of the most unknown and undervalued. Its effects go far beyond the difficulty of making astronomical observations, since according to recent studies, this phenomenon can have serious consequences for people's health [3,10,19]. Another of its negative points, perhaps best known and studied, is the negative impact it has on the ecosystems found under these skies, particularly affecting nocturnal insects [6,8,17]. To all this we must add the energy expenditure involved in so many sources of artificial light. Currently, Spain is the European country that spends the most money per year on public lighting [14]. Much of this energy is obtained from fossil fuels, which exacerbates the problem of greenhouse gases.

1.2 Proposed Solution

Taking all this into account, our work offers a twofold solution: On the one hand, the creation of a physical device to measure the level of real light that is present in a certain place. On the other hand, a platform is offered for the management and storage of the data read by the photometers. There, thanks to the collaboration with the University of Barcelona, users can obtain a global vision of the state of light pollution in various areas, allowing the download and use of the data by scientific, academic organizations or by the public administration. Thanks to this platform, our proposal stands out from other projects previously developed in relation to this same topic [2,5], where expensive devices with a lack of connectivity options are offered. By allowing access to these data, it is intended to increase the population's interest in the negative effects of light pollution, favoring the development of new administrative and scientific measures to mitigate and combat this effect [12].

Additionally, the platform provides a reference value of the natural sky night brightness, via the GAMBONS astronomical model (see Sect. 3). This reference value is compared with the actual measurement to account for the natural variability of the night sky brightness, due for instance to the presence/absence of the Milky Way or the different aerosol composition of the Earth's atmosphere.

2 Free Dark Sky Meter

2.1 Principles

The FreeDSM photometer is an open hardware and open software solution for citizen light pollution monitoring based on IoT technologies. It has been created under the idea of being a low-cost (around 20–30 euros), easy-to-build and easy-to-use device, so everyone with minimum knowledge about soldering or electronics can make their own. All the source code of the firmware used by the device will be available in a GitHub repository under a CC BY-NC-SA license once the project is finalized, so that everyone can access it to make modifications that are interesting to them. All information regarding assembly, wiring and initial configuration, as well as a basic instruction manual, will be available in a public Wiki associated with this repository.

2.2 Components and Design

It is based on an ESP32-C3 micro-controller, a small, cheap and low-consumption but powerful device. To measure the ambient light of the night sky, the FreeDSM includes a TSL2591 [1] optical light sensor, a very sensitive component with a wavelength range of 300–1000 nm and a dynamic range of 600,000,000:1, capable of detecting light from 188 uLux up to 88,000 lx. Thanks to the use of a double-side breadboard, additional sensors can be incorporated, allowing the users to expand the device capabilities or implementing new features to measure (See Fig. 1). It includes a Joystick controller and a OLED 0.96' display at the bottom to show current readings. Thanks to that, and the use of a 3200 mAh USB-C rechargeable battery, the device can also be used as a portable tool to take in-place readings, with an autonomy of 22 h of continuous use. The approximate battery level will appear on the screen to give the user some feedback. For this calculation, the level read by an analog pin connected to the battery through a simple voltage divider will be taken.

Finally, it should be noted that in the current state of the project, work is being done to include a temperature and humidity sensor model AHT21 that allows knowing the environmental situation of the device. This information could be of interest in the future (not yet defined) calibration process.

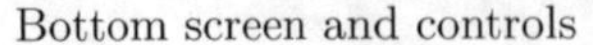

Bottom screen and controls

Breadboard and TSL2591

Fig. 1. Interior of a FreeDSM photometer

In it's current state, the device is designated to be a DIY device, using well-known and popular hardware, some 3D-printable pieces and a segment of

PVC pipe as weatherproof outer casing. With the whole case, the device reaches 13.5 cm high. The diameter of the pipe that has been used is 67 mm, with a thickness of 4 mm. At the top of it, there is a concave 60° PMMA plastic lens, but a new model is under development, replacing this lens with a 15° conical one.

2.3 Software and Connectivity

Regarding the firmware, the device's logic is implement through Tasmota, an open-source framework for smart-home devices. This software offers a set of drivers and common functionalities for the most popular sensors in the market, simplifying the development process of new features.

When used fixed, the device automatically takes a reading every 5 min, enough time to measure the evolution of the night sky without consuming too much energy. Although a reading every minute would be more accurate, the consumption of the device would be considerably higher, so the team considered the current period sufficient. As this is a work still in development, this figure could change over time. From this reading, the values of Infrared, Full Spectrum, Illuminance and Astronomical Magnitudes (mag/arcsec2) are extracted. These values are sent by HTTP to our platform in the appropriate format in case of having internet. Otherwise, they are temporarily stored in the flash memory of the device, until the user accesses and retrieves them through the configuration web page hosted by the device. By using HTTP as the sending method, we leave the MQTT protocol [9] free so that interested users can connect FreeDSM to their private home assistant servers for further integration.

Although the chip has Bluetooth connectivity, it has been decided to implement all connectivity with the device through the WiFi module. A version with LoRaWAN [16] connectivity is also contemplated, but it has not been added to the final design yet.

2.4 Testing and Preliminary Results

For the calibration, an approach focused on neural networks is sought, so having an adequate dataset is essential. The first tests carried out were carried out by placing the FreeDSM next to a SQM-LU-DL [15] for a month, measuring with different configurations for the sensor. One of the comparisons made for one of the nights can be seen below (See Fig. 2). The SQM-LU-DL photometer was selected because of its sensor, a TSL2561 optical light sensor (analog), has been used historically in various sky brightness measurement devices. FreeDSM, however, is equipped with a digital sensor of illuminance TSL2591 (digital, 16bit) which uses two photodiodes (infrared radiation and both infrared and visible spectrum). These independent measurements can be subtracted, easily eliminating the influence of infrared radiation without any filter (like TSL237 devices) decreasing the cost of the device, keeping results very accurate and comparable to analog sensors [20]. In parallel to this development, that was started in 2021 and was initially based on a Raspberry and this same sensor, we have seen

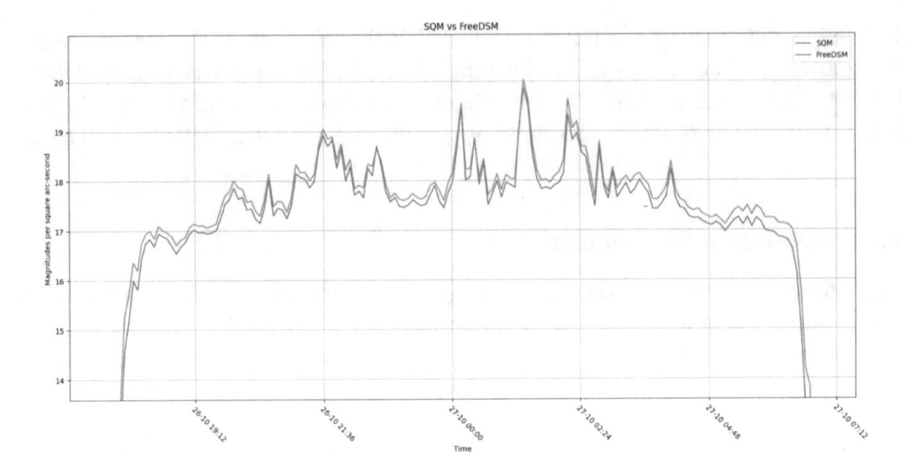

Fig. 2. Comparison between FreeDSM and SQM-LU-DL (October 10th, 2022)

that it is being used in numerous projects that already incorporate recent low-cost communication technologies such as LoRa [7] or large-scale sky brightness measurement campaigns [5], which further supports our choice.

As a note, the approach to calibration using artificial intelligence is still future work, so neither its architecture nor its specifications have been defined. It will be carried out at the end of the project life as an alternative of the calibration coefficient that we are applying right now with good results.

3 Dark Sky Platform

3.1 Architecture

The other element proposed in this work consists in the creation of an open platform in the cloud to support the photometers, allowing the management, storage and manipulation of the readings made by the devices. For the design of this platform, we have chosen to use FIWARE [4], an open framework for IoT solutions that seeks to standardize certain common elements in this type of system. FIWARE defines a set of customizable modules that communicate via REST APIs to perform specific tasks, allowing the users a certain level of escalability and adaptability based on their needs. The data flow within the project is as follows (See Fig. 3):

- **IoT Agent**: FreeDSMs are connected to the platform via HTTP using the UltraLight message format. The component in charge of managing this communication and storing the received metadata of the registered devices into the system is called IoT Agent[1].

[1] IoT Agent (Ultralight) - https://fiware-tutorials.readthedocs.io/en/stable/iot-agent/index.html.

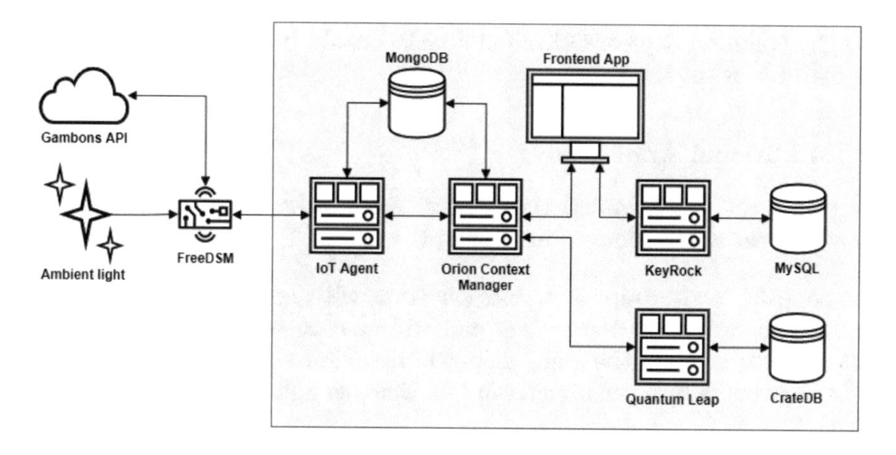

Fig. 3. Platform architecture using FIWARE

- **Orion Context Manager**: The readings made by the photometers are procesed by the Orion Context Manager module[2]. It stores in a separate database the last values read by a device and manage the communications between the other system components.
- **Quantum Leap**: Thanks to the Quantum Leap[3] subscription system, the data read with any interest to users is stored in a historical database, allowing subsequent consultations by time ranges.
- **KeyRock**: Users can register in the system and associate devices to their account using the KeyRock module[4], which offers both authentication and authorization to perform certain operations within the platform.

Fiware has been chosen as the technology for the backend due to its versatility and ease of use. Fiware establishes a series of standards that ensure that all IoT systems implemented on top of it function in the same way, regardless of the area in which they specialize. This is a very important feature to take into account for the solution presented here, since it allows the integration of the platform with other Fiware-based systems that may be developed in the future.

Additionally, the photometers are connected to the GAMBONS[5] astronomical model [13], developed by the University of Barcelona[6] using data from the Gaia space mission [18], to request the theoretical natural brightness level present in the same area in which the device is located. In this way, with the theoretical value and the real value, FreeDSMs can calculate an approximation of the ambi-

[2] Orion Api Reference - https://github.com/telefonicaid/fiware-orion/blob/master/doc/manuals/orion-api.md.
[3] Quantum Leap - https://quantumleap.readthedocs.io/en/latest/.
[4] KeyRock - https://fiware-idm.readthedocs.io/en/latest/.
[5] GAMBONS - https://gambons.fqa.ub.edu/.
[6] Universitat de Barcelona - https://web.ub.edu/inici.

ent light pollution degree. Other public APIs could be used as well to include new features in the future.

3.2 Additional Applications

Along with the platform and the devices, the solution proposed here includes two additional applications of interest to the users.

- **FreeDSM web map**: A web application with an interactive map where users can register in the system and add or manage the devices they own. Navigating through the map, users will be able to access and download the readings made by a photometer in real time, as well as view the metadata of that FreeDSM.
- **FreeDSM flasher**: A desktop application with which to perform the initial configuration of the devices. It allows less experienced users to install all the necessary software on the ESP32 chips in the easiest possible way through its simple graphical interface, thus eliminating the need for any knowledge of electronics or programming.

4 Conclusions

FreeDSM is a low cost photometer (20–30 euros) based on IoT technologies and developed with open software and hardware that helps to quantify the degree of light pollution present in a particular area. Being focused on citizen science projects, it has been designed as a tool that is easy to assemble and use. As main component, FreeDSM uses an ESP32-C3 microcontroller, a powerful low-cost device in which a TSL2591 light sensor is integrated. Along with the device, an open platform is offered for the storage and management of data by users.

Although the proposed solution is still under development, several devices have been deployed with the aim of starting outreach campaigns that raise concerns about the increase in light pollution and its effects into the public eye. As an example of this, some collaborations have been created with high school teachers to give electronics workshops based on the construction of FreeDSMs. Some of the work that is being carried out parallel to this consists of the inclusion of additional sensors, such as environmental quality, noise or even salinity sensors, to make FreeDSM a device adaptable to different environments and requirements.

Acknowledgements. This work was funded by the Spanish MCIN/AEI/10.13039/501100011033 and European Union Next Generation EU/PRTR through grants PDC2021-121059-C21/C22 and the Galician Regional Government, Xunta de Galicia, through grants ED431B 2021/36 and ED431G 2019/01.

References

1. Alarcon, M.R., Puig-Subirá, M., Serra-Ricart, M., Lemes-Perera, S., Mallorquín, M., López, C.: SG-WAS: a new wireless autonomous night sky brightness sensor. Sensors **21**(16) (2021). https://doi.org/10.3390/s21165590, https://www.mdpi.com/1424-8220/21/16/5590
2. Barái, S., Tapia, C.E., Zamorano, J.: Absolute radiometric calibration of TESS-W and SQM night sky brightness sensors. Sensors **19**(6) (2019). https://doi.org/10.3390/s19061336, https://www.mdpi.com/1424-8220/19/6/1336
3. Chepesiuk, R.: Missing the dark: health effects of light pollution. Environ. Health Perspect. **117**(1), A20–A27 (2009)
4. Cirillo, F., Solmaz, G., Berz, E.L., Bauer, M., Cheng, B., Kovacs, E.: A standard-based open source IoT platform: FIWARE. IEEE Internet Things Mag. **2**(3), 12–18 (2019). https://doi.org/10.1109/IOTM.0001.1800022
5. Karpinska, D., Kunz, M.: Device for automatic measurement of light pollution of the night sky. Nat. Sci. Rep. **12** (2022). https://doi.org/10.1038/s41598-022-20624-7, https://www.nature.com/articles/s41598-022-20624-7
6. Eisenbeis, G., Hänel, A., McDonnell, M., Hahs, A., Breuste, J.: Light pollution and the impact of artificial night lighting on insects. Ecol. Cities Towns 243–263 (2009)
7. Erwinski, K., Karpinska, D., Kunz, M., Paprocki, M., Czokow, J.: An autonomous city-wide light pollution measurement network system using LoRa wireless communication. Sensors **23**(11) (2023). https://doi.org/10.3390/s23115084, https://www.mdpi.com/1424-8220/23/11/5084
8. Firebaugh, A., Haynes, K.J.: Light pollution may create demographic traps for nocturnal insects. Basic Appl. Ecol. **34**, 118–125 (2019)
9. Hillar, G.C.: MQTT Essentials-A Lightweight IoT Protocol. Packt Publishing Ltd, Birmingham (2017)
10. Kumar, P., Ashawat, M.S., Pandit, V., Sharma, D.K.: Artificial light pollution at night: a risk for normal circadian rhythm and physiological functions in humans. Curr. Environ. Eng. **6**(2), 111–125 (2019). https://doi.org/10.2174/2212717806666190619120211, https://www.ingentaconnect.com/content/ben/cee/2019/00000006/00000002/art00004
11. Kyba, C.C.M., et al.: Artificially lit surface of earth at night increasing in radiance and extent. Sci. Adv. **3**(11), e1701528 (2017). https://doi.org/10.1126/sciadv.1701528, https://www.science.org/doi/abs/10.1126/sciadv.1701528
12. Mander, S., Alam, F., Lovreglio, R., Ooi, M.: How to measure light pollution—a systematic review of methods and applications. Sustain. Cities Soc. **92** (2023). https://doi.org/10.1016/j.scs.2023.104465, https://www.sciencedirect.com/science/article/pii/S2210670723000768
13. Masana, E., Carrasco, J.M., Bará, S., Ribas, S.J.: A multiband map of the natural night sky brightness including Gaia and Hipparcos integrated starlight. Mon. Not. R. Astron. Soc. **501**(4), 5443–5456 (2020). https://doi.org/10.1093/mnras/staa4005, https://doi.org/10.1093/mnras/staa4005
14. Miguel, A.S.D., Zamorano, J.: Light pollution in Spain: a European perspective. In: Diego, J., Goicoechea, L., Gonzalez-Serrano, J., Gorgas, J. (eds.) Highlights of Spanish Astrophysics V, pp. 535–535. Astrophysics and Space Science Proceedings. Springer, Berlin, Heidelberg (2010). https://doi.org/10.1007/978-3-642-11250-8_164

15. Nievas Rosillo, M., Zamorano Calvo, J.: PySQM the UCM open source software to read, plot and store data from SQM photometers. DOCTA Complutense (2014). https://hdl.handle.net/20.500.14352/41578
16. Osorio, A., Calle, M., Soto, J.D., Candelo-Becerra, J.E.: Routing in lorawan: overview and challenges. IEEE Commun. Mag. **58**(6), 72–76 (2020). https://doi.org/10.1109/MCOM.001.2000053
17. Owens, A.C., Cochard, P., Durrant, J., Farnworth, B., Perkin, E.K., Seymoure, B.: Light pollution is a driver of insect declines. Biol. Conserv. **241**, 108259 (2020)
18. Riello, M., et al.: Gaia early data release 3-photometric content and validation. Astron. Astrophys. **649**, A3 (2021)
19. Spivey, A.: Light pollution: light at night and breast cancer risk worldwide. Environ. Health Perspect. **118**(12), A525–A525 (2010)
20. Zhang, Y., Wijeratne, L.O.H., Talebi, S., Lary, D.J.: Machine learning for light sensor calibration. Sensors **21**(18) (2021). https://doi.org/10.3390/s21186259, https://www.mdpi.com/1424-8220/21/18/6259

NITRO: A Gadget to Transform Standard Rollators into Smart Rollators for Monitoring User Conditions

Joaquin Ballesteros[1](✉), Manuel Fernandez-Carmona[2], and Cristina Urdiales[2]

[1] ITIS Software, Universidad de Málaga, Málaga, Spain
jballesteros@uma.es

[2] Ingeniería de Sistemas Integrados Group, University of Málaga, Málaga, Spain
{mfcarmona, acurdiales}@uma.es

Abstract. The goal of using a rollator is to provide physical stability, support, and increase independence while walking or performing daily activities. The decision to use a rollator is often made in consultation with healthcare professionals, who assess the individual's mobility limitations and recommend the appropriate assistive device. Rollators can be made "smart" via the inclusion of sensors and actuators, so that they can monitor user condition. This allows caregivers or family members to remotely monitor the user's condition, track their activities, and receive alerts or notifications if any issues arise. That is particularly important in care home facilities, where the ratio of users per caregiver is high. This work presents the design and initial test of NITRO, a gadget that tur**N**s a rollator **I**nto a smar**T** **R**Ollator to monitor the user's gait and routines in care home facilities. Two main requirements have been imposed by the caregivers: first, the gadget should be attached (and detached) to any standard rollator frame (four-wheels, three-wheels, two rear wheels), so users can continue using their rollators: and seconds, it can monitor at any time and anywhere while they use their rollators.

Keywords: Smart rollators · Assitive devices · HealthCare 4.0

1 Introduction

The percentage of the world population aged 65 and above is projected to grow by 13.5% in the coming decades, as reported by the authoritative source, Desa 2019 World Population [8]. This significant increase in the elderly population is closely linked to a rise in disabilities and dependence among older individuals, posing challenges to an already strained healthcare system.

To address this issue, it becomes imperative to prioritize and extend the autonomy of older people. Recognizing this urgency, the World Health Organization (WHO) highlighted the importance of this matter in its Global Strategy and Action Plan on Aging and Health under strategic objective 2.1 [16]. WHO

J. Bravo and G. Urzáiz (Eds.): UCAmI 2023, LNNS 841, pp. 25–35, 2023.
https://doi.org/10.1007/978-3-031-48590-9_3

emphasizes the potential of assistive devices in enabling older individuals to maintain control over their lives, provided these devices are tailored to their specific needs and adapted to their unique environments.

To preserve autonomy, personalized solutions are crucial to avoid problems such as disuse syndrome, loss of residual skills, or abandonment of assistive technology. Individualized assistive devices are pivotal in empowering older individuals and ensuring their independence.

Among various types of assistive devices, those designed to enhance mobility are significant contributors to promoting autonomy. Mobility loss adversely affects several Activities of Daily Living (ADL), including leisure activities engagement, social interactions, and utilization of public transportation [7]. Canes, rollators, and wheelchairs are the primary categories of mobility assistive devices [26], each catering to specific user requirements. This work focuses on a subset of assistive devices used when the user needs support for loading and balancing while walking, the rollators.

While many mobility of these rollators offer some mechanical adaptability, these conventional solutions have limitations. Manual adjustments are often required, and mechanical adaptations can only go so far in meeting individual needs. Alternatively, "smart" assistive devices have emerged, incorporating sensors and actuators to facilitate seamless interaction between the user and the device [2,3,11].

These smart rollators, equipped with sensors, hold significant potential in monitoring users' activity, biomechanical parameters, and other relevant data, providing valuable insights into their condition and lifestyle. This data-driven approach enables better-informed decisions and more effective support for older adults, enhancing their overall quality of life. While other devices such as wearables may be forgotten or can be rejected by users, our approach relies on sensoring a device that has already been accepted and it is part of their routine. As a consequence, it is unlikely that users forget to use their walker, resulting on higher degrees of adherence to the monitoring process.

Hence, by acknowledging the projected growth of the elderly population and its potential challenges, we must confidently prioritize extending older people's autonomy. Adopting personalized and smart assistive devices, particularly in mobility, will empower older individuals to lead fulfilling, independent lives while easing the burden on healthcare resources. This proactive approach aligns with the WHO's vision for ageing and health and represents a promising step forward in addressing the needs of an ageing global population.

The rest of the paper is organized as follows: Sect. 2 presents the state of the art in smart rollators and motivates the creation of NITRO. Section 3 describes NITRO, a cyber-physical system that transforms a regular rollator into a smart rollator. Section 4 describes the test to validate that monitoring can be performed. Lastly, Sect. 5 outlines the conclusion and future lines.

2 Related Work

The field of smart rollators has seen remarkable advancements in user interfaces, particularly in how these devices adapt to the individual needs of users. Initially, early smart rollators operated with conventional input devices like touchpads or joysticks to determine user intentions. However, the interfaces rapidly evolved to become more intuitive and seamless by leveraging onboard sensors to infer meanings from the physical interaction between the user and the rollator. Authors in [25], reviewed recent approaches, and conclude that unanimously the interfaces on smart rollators utilize on-board sensors, such as force sensors [4,19], force and torque sensors [13], or load cells [20]. Additionally, some platforms incorporate voice interaction systems, allowing users to express their intentions verbally [15].

Another field of research in intelligent rollators is the operation modes tailored to users' needs. One common mode is the Shared-control mode, found in 40% of the reviewed works in [25]. This mode provides generic assistance to users, primarily assisting them in steering to the correct orientation and maintaining secure distances between the rollator and the user or obstacles [1,12,18].

However, a growing focus on the rollator area is personalizing assistance to accommodate users with varying disability profiles. For example, some studies use models to forecast human motion, enabling the platform to adapt its movement to the user's gait while considering constraints such as separation distance and weight bearing [6]. Admittance controllers continuously adjust support based on long-term user performance models, accounting for factors like physical fatigue or velocity [5,13]. Reinforcement learning techniques are also employed to optimize user contribution weight to control, maximizing safety (distance to nearby obstacles) and trajectory smoothness [28].

Additionally, smart rollators often operate in an autonomous-human mode, monitoring users' gait and biomechanics to assess their condition and trends. Various sensors, such as IMUs, encoders, LIDAR, RGB-D cameras, and force sensors, collect data on walking speed, distance travelled, and variability between left and right walking patterns, providing valuable insights into users' conditions. This data can even predict medical assessments like the Tinetti Mobility Assessment [22].

Furthermore, monitorization can provide non-physical assistance, such as visual or acoustic feedback, to improve gait abnormalities in users with Parkinson's disease or elderly gait issues [14,27]. However, the general approach of these solutions is to take a new rollator and transform them into smart. They fit their solutions to work with specific models. Some solutions cut the handlebars to include force or torque sensors [19,20]. Other soldering sensors into the handlebars [10]. In the extreme, they build an utterly new rollator frame [13]. These approaches work pretty well during rehabilitation because users do not have their own rollators or have had them for a short time.

Nevertheless, this approach does not work correctly at care home facilities, where the users have their rollators for months or even years and resist to change to a new one. This is the primary barrier that the authors have found using Walk-IT [11] at care home facilities, and the reason to create NITRO. In [21], authors investigated the levels and factors influencing the abandonment of assistive products and concluded that walkers (and hence rollators) have a higher abandonment ratio. The simultaneous use of multiple assistive products increases it notably.

3 NITRO Platform

The NITRO platform is bassically composed by three main elements: a computing unit, a set of sensors and the software to process those sensor's data in the computing unit. In this section we will describe them in detail, showing our current implementation.

3.1 Hardware Architecture

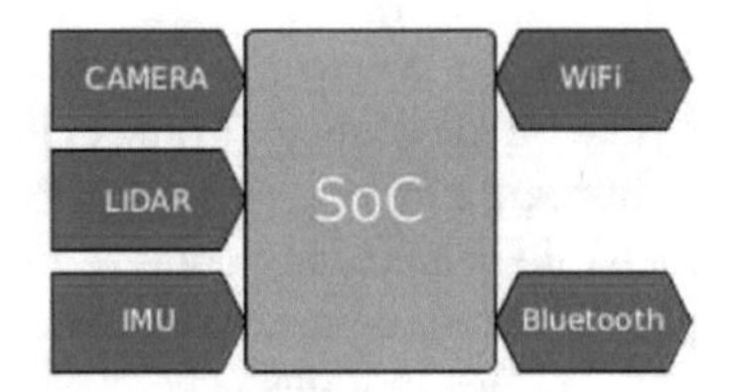

Fig. 1. NITRO hardware architecture.

NITRO hardware architecture is built around off-the-shelf components with price and reproducibility in mind. Where possible, standard buses and interfaces are favoured. This philosophy aims to develop an easily adaptable and maintained system while keeping minimum costs.

The basic structure of the hardware architecture is summarized in Fig. 1. The core of the NITRO hardware is a System on a Chip (SoC) computer. SoCs have reasonable power requirements for a portable system and integrate most of the communication and hardware buses needed. Specifically, wireless communications and high-speed buses (i.e. USB 3.0) are the most relevant for our system for communication and storage. SoCs also have another advantage; their architectures are usually x64 or ARM64 based. These architectures have extensive support in terms of software libraries and hardware compatibility.

Besides communication and general user data management, NITRO hardware must manage sensors for user monitoring, i.e. implement drivers that abstract from specific implementations and offer the same sensor data structures no matter their origin. According to previous works [11], these sensors need to cover two essential tasks for user monitoring in rollators: rollator odometry and leg detection. These two tasks enable stability and gait analysis in rollators.

Leg detection can be achieved with a Laser Imaging Detection and Ranging (LIDAR) sensor attached to the rollator (in the NITRO system). Legs will be close to the rollator and limited to a known area, simplifying the detection process. LIDAR will provide information about how legs move concerning the rollator frame.

Rollator odometry is needed to translate this relative leg motion to absolute coordinates fully. This could be obtained from wheel encoders, but such an approach requires additional adjustment to the rollator and is prone to errors, as sometimes rollators are not pushed but slid or moved without touching the floor.

Instead, we propose using Visual Inertial Odometry (VIO) algorithms [23]. These algorithms fuse speed and acceleration data from Inertial Measurement Units (IMU) with optical flow obtained from a stereo camera to estimate real odometry without using encoders. This approach is far less invasive with the rollator, easing its installation on any device. Visual data is focused on the floor and does not leave the local unit, so it is safe from a privacy point of view.

NITRO platform implementation has an Nvidia Jetson Nano Developer Kit[1] as SoC. It has no wireless communications interface, so an Intel 8265NGW was added for WiFi and Bluetooth support. Other communication solutions, such as 5G modems, can be used for mobile networks, but it does not included since the focus of the work is to be deployed in care home facilities within Wi-Fi access. Bluetooth is used to connect other optional devices, such as wearables to manage different user profiles, while WiFi is only necessary for intermittent data transmision. Recorded data is only transmitted under request or periodically, if it's in range of a local connection, i.e. while charging at home.

Apart from large computing power, this ARM64 platform offers the possibility to run multiple neural networks in parallel to improve our current gait and stability algorithms. Its sensor equipment is composed by (see Fig. 2):

- IMU: Wit-motion BWT901CL.[2]
- Camera: Intel Realsense D435.[3]
- LIDAR: Slamtec RPLidar A1.[4]

The overall system is fixed to the rollator structure using pipe hangers on the main rollator structure. The only positioning requirement is that the LIDAR faces user legs and camera points forward. Potentially, custom hangers could be designed to adapt NITRO to less ordinary rollator frames.

3.2 Software Architecture

Figure 3 shows the main entities in NITRO architecture. It comprises the following components:

[1] https://developer.nvidia.com/embedded/jetson-nano-developer-kit.
[2] https://www.wit-motion.com/9-axis/witmotion-bluetooth-2-0-mult.html.
[3] https://www.intelrealsense.com/depth-camera-d435/.
[4] https://www.slamtec.ai/home/rplidar_a1/.

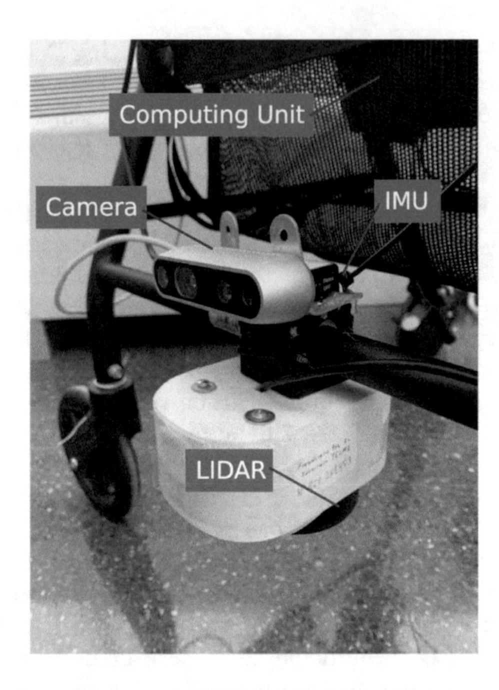

Fig. 2. NITRO platform components: LIDAR (below), stereo camera (front) and IMU (behind the camera).

- **Condition Monitoring Modules**: This module within each rollator processes sensor data, including gait patterns, walking speed and distance travelled. It continuously monitors the user's condition and generates relevant information for dissemination.
- **Supervisor entity**: Users, such as healthcare providers, caregivers, family members, or even the patient are able to subscribe to the rollators' condition monitoring modules. These subscribers receive real-time updates, enabling them to monitor users' well-being remotely.
- **ROS2 Middleware**: The ROS2 middleware serves as the communication backbone, enabling seamless data exchange between rollators and external entities. It supports a rich variety of Quality of Service policies to manage comunication efficient or real-time messaging [17] with low latency and high reliability. It can be configured to work on wifi network, but also on mobile network if needed.

Using ROS2 with Data Distribution Service (DDS) communication, specifically the publish-subscribe paradigm, offers several significant advantages in NITRO. The publish-subscribe enables seamless and efficient data distribution among interconnected rollators and supervisors. By utilizing DDS, the architecture benefits from its high scalability, robustness, and low-latency communication. It also allows for decentralized and asynchronous communication, enabling multiple rollators to broadcast their users' condition data concurrently. This

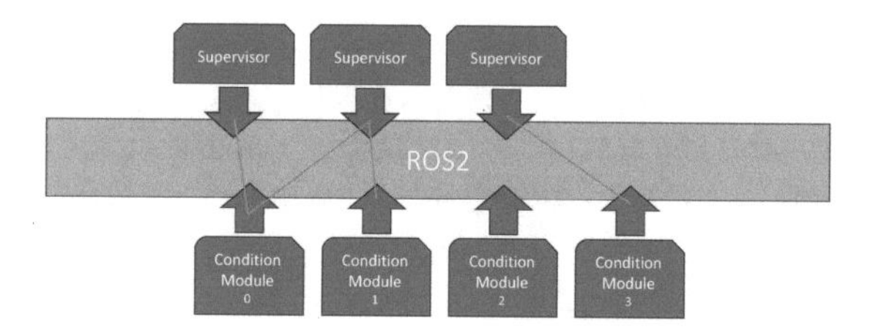

Fig. 3. NITRO software architecture. Condition modules process information from rollators while supervisors gather this information. ROS2 is the communications backbone.

distributed nature ensures that the system can easily scale to accommodate a growing network of rollators and supervisors, making it adaptable to various care home facilities.

Moreover, DDS's built-in mechanisms for data filtering, topic discovery, and data history management enhance the system's performance and reliability. Rollators can publish only relevant condition data, reducing unnecessary network traffic and conserving bandwidth. Meanwhile, subscribers can discover and select specific topics of interest, receiving only the information they require, ensuring data accuracy and minimizing processing overhead. That is represented in the orange lines in Fig. 3.

4 Preliminary Tests

NITRO needs two main functionalities for monitoring users: leg detection (using LIDAR) and odometry (using a stereo camera and an IMU). Combining both, some spatiotemporal gait parameters and fall predictors can be estimated online. The leg detection using LIDAR where broadly investigated in Walk-IT rollator [11]. However, while that rollator used encoders to calculate the odometry, here they have been removed as motivated in Sect. 3.1, and now NITRO uses a stereo camera and an IMU (see Fig. 2) to calculate the odometry. The tests are focused on proving that visual inertial odometry can be used to detect, accurately, the rollator movements.

We have built ROS2 (humble version), Basalt Visual Inertial Odometry [24] and Intel RealSense D400 cameras from source in an Ubuntu 20.04 deployed in a Jetson Nano Developer Kit. An ISO image and some useful guidelines to survive to build from source for Jetson Nano can be found in our GitHub repository[5].

Two types of tests have been carried out. The first focuses on the accuracy in the velocity, and the second on the position. The first test uses the same idea as the well-known ten-meter test; a user walks at a constant speed for ten meters, and a visual mark in the 2 and 8 m is placed. An observer used a clock to measure the time spent walking through the six meters between the 8 and 2 m marks. A user performs 20 tests at different velocities. Figure 4 shows the measured vs

[5] https://github.com/TaISLab/WalKit.

the estimated by the VIO algorithm. It can be observed that the higher the speed, the worse the estimation. That is expected since the VIO algorithm uses marks in the scene and the IMU to estimate the velocity. If you move slowly, the algorithm has more chances to find spots to improve its estimations.

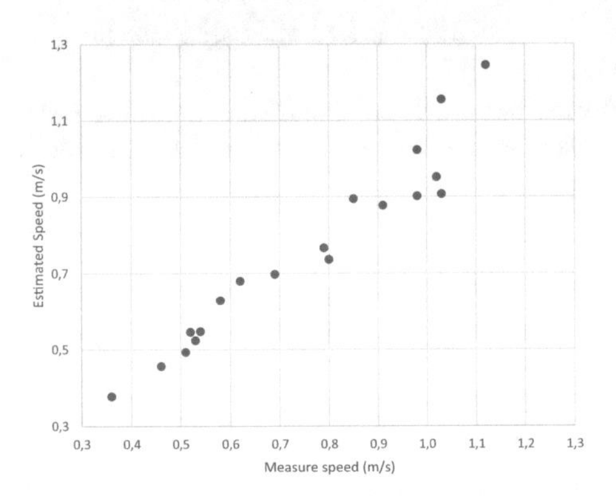

Fig. 4. Velocity estimation using VIO.

The second test focuses on measuring errors in the position estimation. Two subtests have been carried out 20 times each. The first one measures the linear displacement error; the user moves forward ten meters at different locations. The second subtest focuses on performing loops (which include linear but also angular displacements). The user starts and ends in the same position and moves freely for 30 s. Figure 5 shows two boxplots for the errors obtained (differences in meter between the estimated final position and the real ones). It can be observed how the angular displacements impact the accuracy. That happens because the linear displacement involves fewer changes in the scene (from a camera perspective) than the angular displacements.

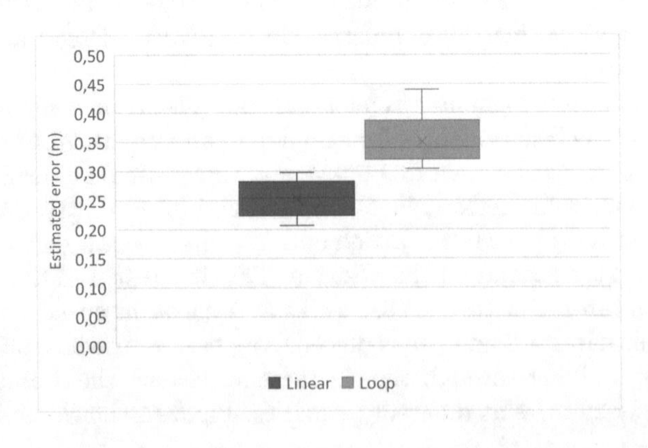

Fig. 5. Position estimation error using VIO.

5 Conclusions and Future Work

NITRO proposes a new gadget integrating a stereo camera to perform Visual Inertial Odometry and a LIDAR to analyze the user condition. It is designed to transform any standard rollator into a smart rollator without significant modifications. It has been built around off-the-shelf components with price and reproducibility in mind. In addition, the codes in NITRO are open source and can be made for different architecture if the SoC is changed.

The proposed architecture uses as backbone ROS2 with DDS using the publish-subscribe model. It can efficiently exchange real-time data among interconnected rollators and supervisors, enabling continuous user monitoring. The decentralized and scalable nature of the architecture ensures adaptability to various care home facilities. The error obtained in the odometry suggests that the system will adequately monitor spatiotemporal gait parameters, such as walking speed or stride length, since the rollator users move relativity slowly compared with the tested users in this paper.

NITRO can be used with other wearable devices to support additional functionalities, like stress analysis studied by authors in [9] or allow multiple user profiles identified by those same wearables.

Future research will focus on implementing and testing the proposed architecture in care home facilities to validate its usability and performance. Conducting questionnaires and user feedback surveys will be essential to gather insights from users and caregivers, validating the effectiveness and usefulness of the system in meeting their needs.

Additionally, efforts will be directed towards optimizing the architecture's energy consumption. Extending the rollators' battery life will be critical to ensuring prolonged and reliable operation in various settings. By addressing these technical challenges, the system can become even more accessible and practical for everyday use. Furthermore, the GPU capabilities of the new SoC selected open a new area of monitoring improvement by the possibility of including artificial neural networks.

Acknowledgements. This work is partially funded by Ministerio de Ciencia e Innovación, Ref: SAVIA RTI2018- 096701-B-C21, Ref: IRIS PID2021-122812OB-I00 (co-financed by FEDER funds)), Ref: CAMPERO TED2021-131739B-C21, by the project UMA-CEIATECH23-2020 from Plan Propio de Investigación of the University of Málaga and Horizon 2020 Ref: H2020-101017109 (DAEMON). The authors would like to thank to Macrosad Arroyo de La Miel carehome for their support on defining the design requirements.

References

1. Andreetto, M., Divan, S., Ferrari, F., Fontanelli, D., Palopoli, L., Prattichizzo, D.: Combining haptic and bang-bang braking actions for passive robotic walker path following. IEEE Trans. Haptics **12**(4), 542–553 (2019)

2. Ballesteros, J., Ayala, I., Caro-Romero, J.R., Amor, M., Fuentes, L.: Evolving dynamic self-adaptation policies of mhealth systems for long-term monitoring. J. Biomed. Inform. **108**, 103494 (2020)
3. Ballesteros, J., Urdiales, C., Martinez, A.B., Tirado, M.: Automatic assessment of a rollator-user's condition during rehabilitation using the i-walker platform. IEEE Trans. Neural Syst. Rehabil. Eng. **25**(11), 2009–2017 (2017)
4. Ballesteros, J., Urdiales, C., Velasco, A.B.M., Ramos-Jiménez, G.: A biomimetical dynamic window approach to navigation for collaborative control. IEEE Trans. Hum.-Mach. Syst. **47**(6), 1123–1133 (2017)
5. Chalvatzaki, G., Papageorgiou, X.S., Maragos, P., Tzafestas, C.S.: User-adaptive human-robot formation control for an intelligent robotic walker using augmented human state estimation and pathological gait characterization. In: 2018 IEEE/RSJ International Conference on Intelligent Robots and Systems (IROS), pp. 6016–6022. IEEE (2018)
6. Chalvatzaki, G., Papageorgiou, X.S., Maragos, P., Tzafestas, C.S.: Learn to adapt to human walking: a model-based reinforcement learning approach for a robotic assistant rollator. IEEE Robot. Autom. Lett. **4**(4), 3774–3781 (2019)
7. Chikaraishi, M.: Mobility of the elderly. In: Zhang, J. (ed.) Life-Oriented Behavioral Research for Urban Policy, pp. 267–291. Springer, Tokyo (2017). https://doi.org/10.1007/978-4-431-56472-0_10
8. DESA, U.: World population prospects 2019: Highlights. New York (US): United Nations Department for Economic and Social Affairs (2019)
9. Díaz-Boladeras, M., et al.: Perceived distress in assisted gait with a four-wheeled rollator under stress induction conditions. Cogent Eng. **10**(1), 2233743 (2023)
10. Fernandez-Carmona, M., Ballesteros, J., Díaz-Boladeras, M., Parra-Llanas, X., Urdiales, C., Gómez-de Gabriel, J.M.: Walk-IT: an open-source modular low-cost smart rollator. Sensors **22**(6), 2086 (2022)
11. Fernandez-Carmona, M., Verdezoto, G., Ballesteros, J., Gómez-de Gabriel, J.M., Urdiales, C.: Smart rollators as a cost-effective solution for personalized assistance healthcare ecosystem in elderly communities. In: Bravo, J., Ochoa, S., Favela, J. (eds.) Proceedings of the International Conference on Ubiquitous Computing and Ambient Intelligence (UCAmI 2022). UCAmI 2022. LNNS, vol. 594, pp. 449–461. Springer, Cham (2023). https://doi.org/10.1007/978-3-031-21333-5_45
12. Ferrari, F., et al.: Human-robot interaction analysis for a smart walker for elderly: the ACANTO interactive guidance system. Int. J. Soc. Robot. **12**(2), 479–492 (2020)
13. Geravand, M., Werner, C., Hauer, K., Peer, A.: An integrated decision making approach for adaptive shared control of mobility assistance robots. Int. J. Soc. Robot. **8**(5), 631–648 (2016)
14. Golembiewski, C., et al..: The effects of a positional feedback device on rollator walker use: a validation study. Assist. Technol. 1–8 (2019)
15. Moustris, G., et al.: The i-walk lightweight assistive rollator: first evaluation study. Front. Robot. AI **8** (2021)
16. Organization, W.H., et al.: Global strategy and action plan on ageing and health. World Health Organization (2017)
17. Puck, L., et al.: Performance evaluation of real-time ROS2 robotic control in a time-synchronized distributed network. In: 2021 IEEE 17th International Conference on Automation Science and Engineering (CASE), pp. 1670–1676 (2021)
18. Ragaja, S., Dinesh, N., Madhuri, V., Parameswaran, A.: Development and clinical evaluation of a posterior active walker for disabled children. J. Intell. Robot. Syst. **97**(1), 47–65 (2020)

19. Sato, W., Tsuchida, Y., Li, P., Hasegawa, T., Yamada, Y., Uchiyama, Y.: Identifying the effects of assistive and resistive guidance on the gait of elderly people using a smart walker. In: 2019 IEEE 16th International Conference on Rehabilitation Robotics (ICORR), pp. 198–203. IEEE (2019)
20. Sierra, M.S.D., Garzón, M., Munera, M., Cifuentes, C.A., et al.: Human–robot–environment interaction interface for smart walker assisted gait: agora walker. Sensors **19**(13), 2897 (2019)
21. Sugawara, A.T., Ramos, V.D., Alfieri, F.M., Battistella, L.R.: Abandonment of assistive products: assessing abandonment levels and factors that impact on it. Disabil. Rehabil. Assist. Technol. **13**(7), 716–723 (2018)
22. Tinetti, M.E.: Performance-oriented assessment of mobility problems in elderly patients. J. Am. Geriatr. Soc. (1986)
23. Usenko, V., Demmel, N., Schubert, D., Stueckler, J., Cremers, D.: Visual-inertial mapping with non-linear factor recovery. IEEE Robot. Autom. Lett. (RA-L) **5**(2), 422–429 (2020). Int. Conference on Intelligent Robotics and Automation (ICRA)
24. Usenko, V., Demmel, N., Schubert, D., Stückler, J., Cremers, D.: Visual-inertial mapping with non-linear factor recovery. IEEE Robot. Autom. Lett. **5**(2), 422–429 (2019)
25. Verdezoto, G., Ballesteros, J., Urdiales, C.: Smart rollators aid devices: current trends and challenges. IEEE Trans. Hum.-Mach. Syst. (2022)
26. Webster, J., Murphy, D.: Atlas of Orthoses and Assistive Devices E-Book. Elsevier Health Sciences, Amsterdam (2017)
27. Wu, H.K., Chen, H.R., Chen, W.Y., Lu, C.F., Tsai, M.W., Yu, C.H.: A novel instrumented walker for individualized visual cue setting for gait training in patients with Parkinson's disease. Assist. Technol. **32**(4), 203–213 (2020)
28. Xu, W., Huang, J., Wang, Y., Tao, C., Cheng, L.: Reinforcement learning-based shared control for walking-aid robot and its experimental verification. Adv. Robot. **29**(22), 1463–1481 (2015)

Dynamic Service Level Agreements and Particle Swarm Optimization Methods for an Efficient Resource Management in 6G Mobile Networks

Borja Bordel(✉), Ramón Alcarria, Tomás Robles, and Miguel Hermoso

Universidad Politécnica de Madrid, Madrid, España
{borja.bordel,ramon.alcarria,tomas.robles}@upm.es,
miguel.hermoso@alumnos.upm.es

Abstract. Future 6G networks are envisioned to provide ultra-massive machine-type communications. But such a huge number of devices implies an extraordinary resource consumption, which surely will exceed the network resources. To mitigate this problem, static optimization algorithms are used to efficiently distribute existing resources. However, this approach presents three basic problems. First, every 6G device must fulfill a different business case. So, some Service Level Agreements may be breached more easily than others. Second, in 6G mobile networks, base stations can increase their resources dynamically, although their operation cost would increase. However, some devices could accept this additional charge. And third, 6G mobile devices should know the actual Service Level Agreement offered by each base station before stablishing the final connection. Therefore, in this paper, we propose a new resource management solution for 6G networks, based on the union of static optimization algorithms and Blockchain-enabled Service Level Agreements, which can be renegotiated dynamically. A transparent Blockchain network allows 6G devices to negotiate their Service Level Agreement with different base stations, before stablishing any connection. The guaranteed Quality-of-Service, the maximum Quality-of-Service, and the tariffication are included in Smart Contracts. Particle swarm optimization algorithms are employed to allocate resources and study the future potential resource distribution. A multilevel optimization scheme is proposed, so we ensure that devices receive resources according to their Service Level Agreement category. Also, an experimental validation based on simulation tools is provided. Results show that the Service Level Agreement fulfillment rate increases by up to 31% compared to equivalent static optimization mechanisms.

Keywords: 6G networks · particle swarm optimization · Blockchain · Service Level Agreement · resource management · Quality of Service

1 Introduction

Future 6G networks are characterized by a collection of different Key Performance Indicators (KPI) [1], describing the Quality-of-Service (QoS) to be achieved, the scenarios to be served, the network structure to be deployed, etc. Among all these KPI, three are the most relevant. Namely:

J. Bravo and G. Urzáiz (Eds.): UCAmI 2023, LNNS 841, pp. 36–47, 2023.
https://doi.org/10.1007/978-3-031-48590-9_4

- Extremely reliable and low-latency communications (ERLLC). In 6G networks, communication delays cannot exceed ten microseconds and reliability must be above 99.999%
- Further-enhanced mobile broadband (FeMBB). 6G devices must be provided with a sustainable capacity of 10 Gpbs and support peaks above one terabit per second.
- Ultra-massive machine-type communications (umMTC). In future 6G networks, base stations must serve without congestion up to ten million devices per square kilometer.

Such an enormous number of devices to be served, with those very hard QoS requirements, implies extraordinary resource consumption. But most critical network resources, such as power and radiofrequency spectrum, are limited and legally restricted [2]. So, in the general case, network resources cannot be increased to meet the demand, and, surely, the total resource consumption will exceed the network resources. In this context, an optimization algorithm to find the most efficient resource distribution among all 6G devices is the only solution to mitigate this problem [3]. But typically, this optimization algorithm is static and runs in one step. Then three basic challenges emerge.

First, every device fulfills a different business case and Service Level Agreement (SLA) [4]. Some services are legally binding and their characteristics (QoS, tariffication…) cannot be changed. Some other services may be "universal" and must be present at every single base station without restriction. On the other hand, different users require a certain SLA (and pay for it). Any change in the level of service provided will definitely cause users to demand a new fee. Therefore, resource optimization cannot be static, but evolve according to the SLA of 6G devices the network is serving at each specific moment.

Second, although network resources are legally restricted, new software-oriented architectures allow 6G base stations to dynamically ask for new resources (if available) and increase their capabilities [5]. But those resources (and the services they support) are usually more expensive to obtain and operate. And may exceed the cost initially recognized in the SLA. But, under certain circumstances, 6G devices and users could be open to modify this initial SLA temporarily, in order to pay an extra fee for new resources. This opens a new market that is relevant for network operators and users. Then, resource management should not be executed in one step. On the contrary, a multistep dynamic negotiation process between 6G networks and devices should be promoted to dynamically update SLA.

And third, and finally, 6G networks are envisioned to provide broad and deep coverage [6]. So, base stations will cover above 99% of the geographical territory. And 6G devices will move with speeds up to one thousand kilometers per hour. Thus, devices could self-distribute within a certain region if the 6G base station would inform about the SLA offered at each moment. And that could help to achieve a more efficient resource distribution. In addition, 6G users must be enabled to reject a service or connection if it does not satisfy them. Then, resource management cannot be centralized, but a distributed procedure.

In this paper, we address this challenge and propose a new resource management scheme, where accountable and transparent Blockchain networks and SmartContracts allow 6G devices and networks to dynamically negotiate the SLA. 6G base stations will offer an individual SLA to each device, calculated based on a Particle Swarm Optimization (PSO) algorithm. PSO will execute at a multistep scheme, so SLA are respectful with the service category (priority or criticality) and flexibility. Tariffication will be negotiated using a game-theory-based approach. And a one-step bidding algorithm will be used to distribute resources if the PSO cannot find a valid distribution.

The rest of the paper is organized as follows: Sect. 2 analyzes the state of the art on resource management solution for 6G networks. Section 3 presents the proposed new technology, including the architecture and algorithms. Section 4 describes the experimental validation and the results, and Sect. 5 concludes the paper.

2 Resource Management in 6G Networks

Resource management is one of the key problems in 6G networks. But most works address this challenge from an Artificial Intelligence perspective. In general, three different approaches can be identified [7]: deep learning, reinforcement learning, and federated learning. Deep learning solutions are typically applied to network slicing scenarios, where resources are split into different applications and/or devices in a quite long-term structure. Applications to the frequency spectrum [8], physical memory [9], or computation hosting [10] have been reported. But this approach is only adequate to address long-term resource distribution. Although transitory bottlenecks or saturation is still possible. On the other hand, reinforcement learning schemes usually focus on very specific problems such as scalability [11] or mobility [12]. But benefits are limited to those objectives, and no other advantage or different compared to deep learning has been reported. While both present the same disadvantage. Finally, federated learning solutions have been reported to address QoS optimization [13], but results can still improve very significantly, as efficiency is around 50% in the most usual case [14].

Nevertheless, traditional optimization algorithms such as swarm intelligence [15], or descendent gradient [16] have also been studied as technologies to allow efficient resource management in a 6G network or between 6G networks and other wireless infrastructures. Although these schemes do not consider any negotiation process or dynamic optimization procedure.

In addition, resource management in multi-band [17] or heterogeneous networks [18] has been thoroughly studied by some authors. Applications to 6G satellite communications [19], 6G subnetworks crowds [20], and high-density Internet-of-Things systems [21] have been reported. Energy-aware algorithms [22], virtual network functions [23], and dedicated protocols [24] to resource management are proposed in order to achieve an efficient resource distribution in future 6G networks. But this approach is based on new radio spectrum distributions, new radiation patterns, and other similar physical-level solutions. And, as explained for deep learning mechanisms, they mitigate but do not prevent the network saturation. More effective technologies are needed.

Due to its great popularity in the last five years, Blockchain technologies have also been applied to 6G resource management. But the works are very scarce. And the typical approach is focused on centralized control and monitoring [25], rather than on distributed negotiation or dynamic adaptation. In this paper we investigate this gap.

3 Efficient Resource Distribution in 6G Networks Using Blockchain and PSO

Figure 1 shows the architecture of the proposed resource management and distribution solution. In a given geographical region $\mathcal{R}$, a 6G subnetwork with B base stations b_i (1) is serving a population of D radio devices d_j (2).

$$\{b_i i = 1, .., B\} \tag{1}$$

$$\left\{d_j i = 1, .., D\right\} \tag{2}$$

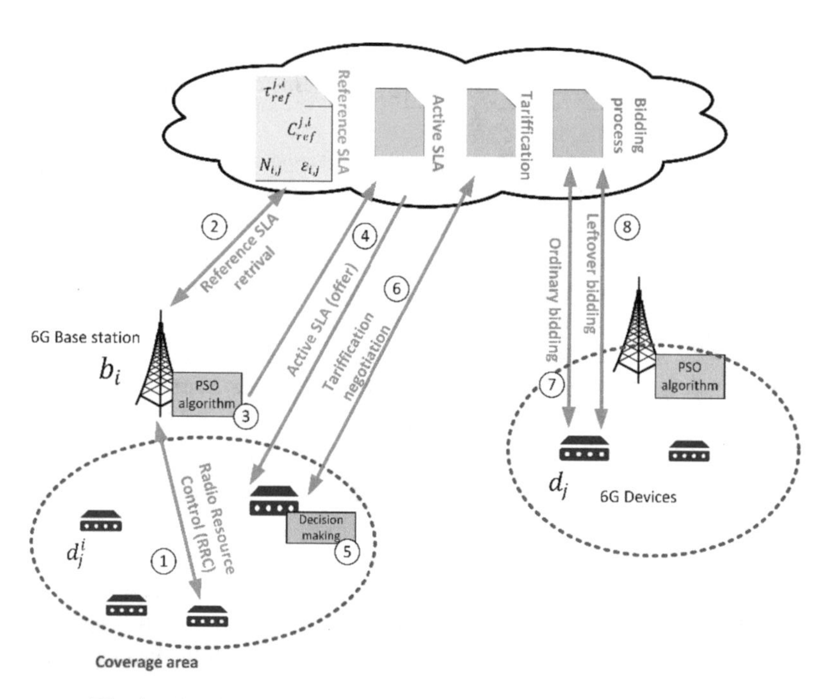

Fig. 1. Architecture for the proposed resource management solution

In 6G networks, as in 5G and 4G systems, handovers and device-to-base-station associations are managed and triggered by the network. But radio devices d_j are mobile, and they could temporarily get closer to a certain base station b_i with no interest in connecting. Thus, to avoid congestion, the SLA negotiation process is triggered by base station b_i every T_i seconds (it is not a continuous procedure). First, using the Radio Resource Control (RRC) protocol, base station b_i identifies the D_{b_i} devices d_j^i within its coverage area ①. Then, for all D_{b_i} devices d_j^i, base station b_i retrieves from the Blockchain network the reference SLA (one per device) ②, where the reference delay $\tau_{ref}^{j,i}$ and bitrate $C_{ref}^{j,i}$ are described, together with the service category $N_{i,j}$ and flexibility $\varepsilon_{i,j}$. For every service category n, the base station b_i defines an optimization problem, to be solved using Particle Swarm Optimization (PSO), in order to distribute resources among devices d_j^i subscribed to services with that category ③. In this context, network resources are represented as a rectangular grid of radio channels and time slots. If the PSO algorithm finds a solution, the new (active) SLA is offered to devices d_j^i though the Blockchain network ④ and can be accepted or rejected ⑤. If accepted, a new negotiation is triggered, based on game theory, to agree the new tariffication ⑥. All this process is performed through the Blockchain to be accountable and binding. But if the PSO algorithm does not find a valid solution, resources are distributed through a one-step bidding procedure ⑦ supported by the Blockchain network, where devices d_j^i offering the highest reward get the resources. Additionally, leftover resources will be auctioned off too ⑧, together with some additional resources, which could be added, if the offer is high enough from the network perspective.

3.1 Efficient Resource Distribution Using PSO: Blockchain-Enable SLA Negotiation

Every T_i seconds, 6G base station b_i sends a “RRC Measurement Control” message to all devices d_j it can reach. Devices d_j, if they want to negotiate with base station b_i must send a “RRC Measurement Report” as response (see Fig. 2-left-). Then, 6G base stations b_i must decide, based on that report, if device d_j is within their coverage area not. Some proposals for the decision-making algorithm can be found in the state of the art [26]. Every device d_j is identified by the International Mobile Equipment Identity (IMEI). In the Blockchain network, a SmartContract (SC) is deployed for each device d_j describing its reference SLA (see Fig. 2-right-). This SC includes two data structures: an unsigned integer representing the IMEI and a struct describing the reference QoS for this device. This struct contains the following data fields:

- Service category, $N_{i,j}$. It indicates the criticality of the priority of this service. It is a natural number. As lower, more critical or priority is the service. Priority services must be satisfied, and resources must be received before any lower priority service.
- Flexibility, $\varepsilon_{i,j}$. Percentage value. Indicates the maximum difference between the reference QoS and the negotiated QoS that the device can accept. It is a hard limit.
- Communication delay, $\tau_{ref}^{j,i}$. This is a time value. We are assuming propagation delays are negligible and queues and
- Effective bitrate, $C_{ref}^{j,i}$. Refers to the capacity at the physical level, including all overheads at any other superior level.

However, network resources (see Fig. 3) are represented as a rectangular grid. One dimension represents time slots (with a duration of T_{slot} seconds), and the second dimension represents the available radio channels (with a bandwidth of Ω_i Hertz). The real bitrate $C^{j,i}$ provided to device d_j^i is proportional to the number $A_{i,j}$ of assigned time-frequency slots (3) and the Signal-to-Noise ratio $SNR_{i,j}$ (we are assuming is constant in this work) through the Shannon-Hartley law.

Identically, real maximum communication delay $\tau^{j,i}$ can be obtained as the maximum time distance between two consecutive assigned slots t_k and t_{k+1} (4).

$$C^{j,i} = \frac{A_{i,j} \cdot T_{slot}}{T_i} \cdot \Omega_i \cdot \log_2\left(1 + SRN_{i,j}\right) \tag{3}$$

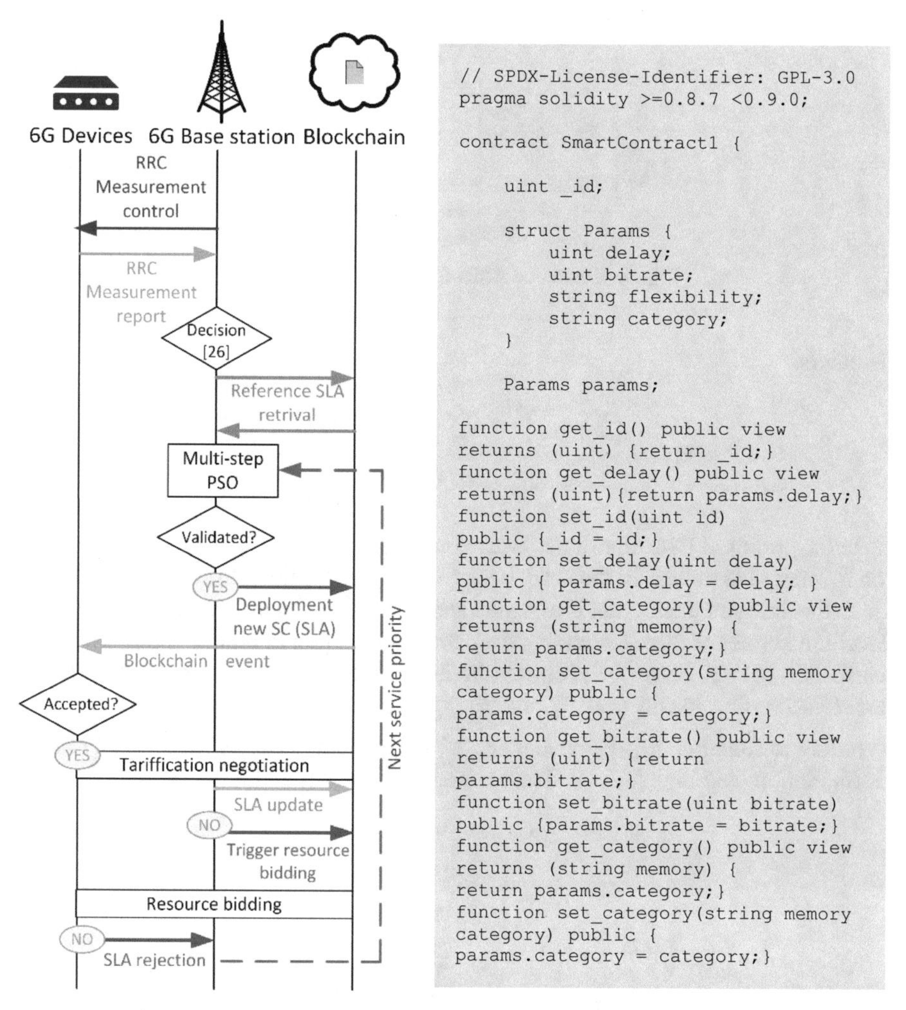

Fig. 2. (Left) Sequence chart (Right) Reference SLA as SC (extract)

$$\tau^{j,i} = \max_{k}\{t_{k+1}, t_k\} \tag{4}$$

With all this information, base station b_i can define a multi-step optimization problem supported by fitness function $\mathcal{F}_{b_i}$ (5). Services are split into groups according to their category, and the optimization process is executed repetitively for increasing category values. In this function, difference between the reference bitrate $C_{ref}^{j,i}$ and the real bitrate $C^{j,i}$, as well the difference between the reference communication delay $\tau_{ref}^{j,i}$ and the real communication delay $\tau^{j,i}$ must be minimized. In addition, the total number of assigned time–frequency slots should be minimum as well. Furthermore, the Heaviside step function $u(\cdot)$ is employed to ensure the provided QoS is never higher than the reference QoS. Parameters $\lambda_{1,2,3}$ are employed to strengthen some indicators against

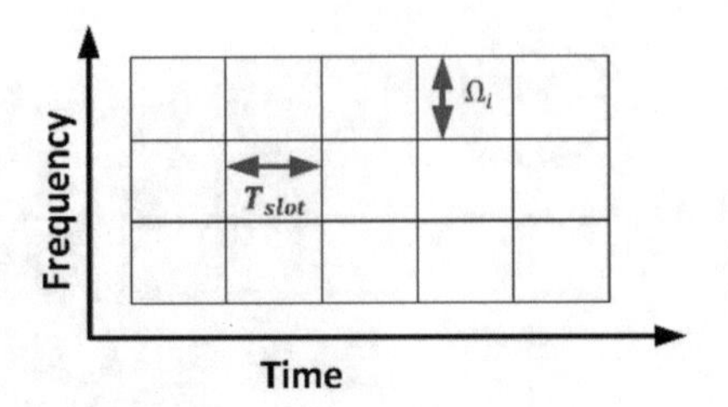

Fig. 3. Resource structure: time–frequency grid

the others.

$$\mathcal{F}_{b_i} = \frac{\left(\sum_j A_{i,j}\right)^{\lambda_1} \cdot C_{ref}^{j,i} - C^{j,i\lambda_2} \cdot \tau_{ref}^{j,i} - \tau^{j,i\lambda_3}}{u\left(C_{ref}^{j,i} - C^{j,i}\right) \cdot u\left(\tau_{ref}^{j,i} - \tau^{j,i}\right)} \tag{5}$$

In the proposed PSO algorithm, every time–frequency slot is a particle, whose position θ_r may take values in the set $\left[0, D_{b_i}\right]$ where each value indicates the devices to which the time–frequency slot is assigned and zero value means "available". Then, PSO algorithm iterates M times, updating the particles' position in order to find an optimum distribution (6) and being k the current iteration and θ_r^{best} the best position of the r-th particle according to function $\mathcal{F}_{b_i}$ and θ_{global}^{best} the best particle's position ever created according to function $\mathcal{F}_{b_i}$. Parameters $z_{(0,1)}^{1,2,3}$ are random numbers following a uniform distribution in the range [0, 1] and parameters $\alpha_{1,2}$ are configuration values.

$$\begin{gathered} \theta_r^k = \theta_r^{k-1} + v_r^k \\ v_r^k = \begin{cases} g_r^k if z_{(0,1)}^3 \\ g_r^k otherwise \end{cases} \\ g_r^k = \alpha_1 \cdot z_{(0,1)}^1 \cdot \left(\theta_r^{best} - \theta_r^{k-1}\right) + \alpha_2 \cdot z_{(0,1)}^2 \cdot \left(\theta_{global}^{best} - \theta_r^{k-1}\right) + w_k \cdot v_r^{k-1} \\ w_k = w_{max} - \frac{(w_{max} - w_{min}) \cdot k}{M} \end{gathered} \tag{6}$$

After the PSO algorithm achieves a solution, it is validated to ensure service flexibility $\varepsilon_{i,j}$ is met. See Algorithm 1. Figure 2 (left) shows the flow chart for the proposed mechanism. If validated, a new SC is deployed describing the offered SLA. This new SC contains a struct with three parameters: the real bitrate $C^{j,i}$, the real communication delay $\tau^{j,i}$ and the tariffication $p^{j,i}$. Besides, it includes functions to retrieve all these values, and a second function to modify the tariffication $p^{j,i}$, , once the negotiation procedure concludes (see Sect. 3.2).

Algorithm 1. PSO solution validation

Input: Proposed QoS $\{C^{j,i}, \tau^{j,i}\}$, reference QoS $\{C^{j,i}_{ref}, \tau^{j,i}_{ref}\}$ and flexibility $\varepsilon_{i,j}$
Output: Validation
Create vector Q with L_Q binary positions
Create integer $L_{current}$ and initialize with value zero
for each device d^i_j **then**
  if $\frac{\|C^{j,i}_{ref}-C^{j,i}\|}{C^{j,i}_{ref}} > \varepsilon_{i,j}$ **or** $\frac{\|\tau^{j,i}_{ref}-\tau^{j,i}\|}{\tau^{j,i}_{ref}} > \varepsilon_{i,j}$ **then**
    return not validated
  end if
end for
return validated

A Blockchain event is then triggered, informing device d^i_j the new SLA is available. Device d^i_j may accept the new SLA, and then the tariffication negotiation is started (see Sect. 3.2). If the new SLA is rejected, assigned resources are unallocated and employed in the next optimization step (services with a higher category -less priority-). Finally, if the PSO solution is not validated, a one-step bidding process is triggered (see Sect. 3.3) to distribute the resources.

3.2 Tariffication Negotiations

When a new SC describing the offered SLA is deployed, if this SLA is accepted by device d^i_j, a tariffication negotiation process is triggered. In this negation, base station b_i can demand any price $p^{j,i}_{bs(s)}$ within the set P_{bs} (7), being p^{min}_{bs} the minimum acceptable price for the 6G provider. Each one of the prices $p^{j,i}_{bs(s)}$ is giving a different profit $a^{j,i}_{bs(s)}$ (of payoff) to the 6G network (depending on the operation cost, type of service, etc.), being a^{min}_{bs} the minimum profit to be achieved (8). On the other hand, device d^i_j can also propose a different price $p^{j,i}_{dev(s)}$ within range P_{dev} (9), causing a payoff (savings) $a^{j,i}_{dev(s)}$ within the range A_{dev} (10). This situation can be understood as a game $\mathcal{G}$, where the 6G base station b_i and device d^i_j are the players (11).

$$P_{bs} = \left\{p^{min}_{bs}, p^{j,i}_{bs(1)}, \ldots, p^{j,i}_{bs(s)}, \ldots \infty\right\} \tag{7}$$

$$A_{bs} = \left\{a^{min}_{bs}, a^{j,i}_{bs(1)}, \ldots, a^{j,i}_{bs(s)}, \ldots \infty\right\} \tag{8}$$

$$P_{dev} = \left\{p^{min}_{dev}, p^{j,i}_{dev(1)}, \ldots, p^{j,i}_{dev(s)}, \ldots \infty\right\} \tag{9}$$

$$A_{dev} = \left\{a^{min}_{dev}, a^{j,i}_{dev(1)}, \ldots, a^{j,i}_{dev(s)}, \ldots \infty\right\} \tag{10}$$

$$\mathcal{G} = [P, A] = \{P_{bs}, A_{bs}; P_{dev}, A_{dev}\} \tag{11}$$

This game is dynamic, so it is necessary to fix a maximum number of rounds for the negotiation process. In this case, we propose that five rounds should be enough to achieve agreement. Besides, device d_j^i is the player doing the first proposal (movement or strategy), so 6G base station b_i can only choose between two economically viable tactics (satisfaction and control tactics implies to give device d_j^i the highest possible payoff and it makes no sense in this scenario):

- Maximum price: Where 6G base station b_i tries to reduce as much as possible the payoff (savings) of device d_j^i and increase profit as much as possible too.
- Perfect agreement: The 6G base station b_i follows a strategy causing a Nash equilibrium, so the game finishes as no player has incentive to do a new offer.

When an agreement is achieved, resulting tariffication $p^{j,i}$ is stored is the corresponding SC describing the new and active SLA.

3.3 Blockchain-Enabled Resource Bidding

Finally, the bidding procedure is quite simple as it follows a one-step scheme. If this mechanism must be triggered, a Blockchain event is published, and all devices d_j^i can ask for resources using a public and transparent bid. Devices d_j^i (all of them have the same category at this stage) are prioritized according to profits each bid generated for the 6G provider. And resources are distributed according to this list until they run out. This entire process is coded in a SC, to which network resources are transferred at the beginning of the biding.

Device d_j^i does a bid of $e_{i,j}$ monetary units, asking for $y_{i,j}$ time-frequency slots (network resources). Devices d_j^i can ask for network resources directly, or ask for a certain QoS, which is later transformed into a certain number of time–frequency slots using previous expressions (3–4). On the other hand, the operation and use of each time-frequency slot has a cost of q_{slot} monetary units. Therefore, it is quite simple to calculate the profits $x_{i,j}$ associated to each bid $e_{i,j}$ (12).

$$x_{i,j} = e_{i,j} - q_{slot} \cdot y_{i,j} \tag{12}$$

3.4 Experimental Validation: Simulation and Results

In order to evaluate the performance of the proposed resource management solution, an experimental validation was conducted. A 6G network was simulated using the NS3 simulator and TAP bridges. The simulation scenario consisted of ten base stations, homogenously distributed in a geographical region of twenty square kilometers. Also, a population of five million devices was looking for mobile services and network resources (to be coherent with future umMTC applications). Devices moved randomly across the entire geographical region.

The proposed resource management solution was implemented in a virtual machine (Ubuntu 22.04 LTS) using Ethereum technologies and the Truffle project. The protocols and all network logic were implemented in the NS3 simulator using C++ code. The virtual machine executing the Blockchain network, and all the SC and the simulation scenario (in the NS3 platform) were connected through TAP bridges.

Each base station in this scenario was provided with a random number of resources (time-frequency slots). And reference SLA were also created randomly, although we ensure that between three and ten different service categories were considered in every single simulation.

All these simulations were executed by a Linux server (Ubuntu 22.04 LTS) with the following hardware characteristics: Dell R540 Rack 2U, 96 GB RAM, two processors Intel Xeon Silver 4114 2.2G, HD 2 TB SATA 7,2 K rpm. Each simulation represented seventy hours of network operation. And every simulation was repeated twelve times to mitigate exogenous variables. Final results are the average of all of these partial simulations.

In the proposed experiment, we measured the SLA fulfillment rate. The QoS provided to each single device was continuously monitored (through the SC deployed in the Blockchain Ethereum network), as well as the reference SLA. The percentage of devices for which the active SLA and the reference SLA are equivalent is calculated at each time instant. These data are analyzed using statistical instruments such as boxplots.

On the other hand, the same experiment was repeated again, but on this occasion, resources were distributed using a static one-step optimization algorithm. To obtain comparable results, we employed the PSO algorithm (6). As before, the SLA fulfillment rate was monitored and analyzed using statistical instruments.

To conclude if a significant improvement was achieved with the proposed resource management solution, a Mann-Whitney U test was used to compare both techniques and distributions. Figure 4 shows the obtained results in boxplots. Future works will explore new metrics in terms of computation time or memory requirements.

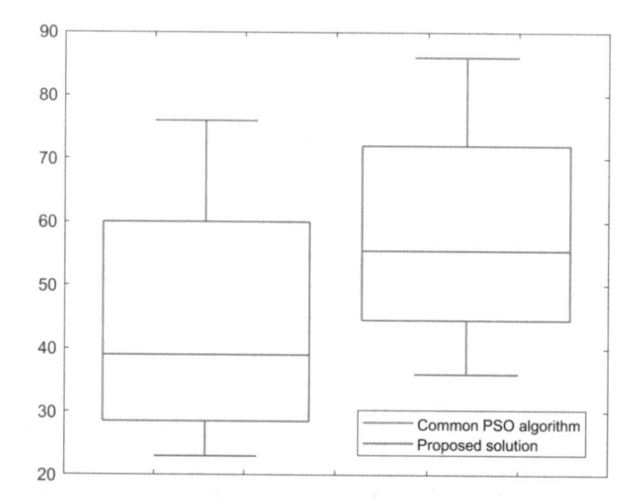

Fig. 4. Experimental results

As can be seen, boxplots overlap, but all relevant values (minimum, average, and maximum) are higher for the proposed resource management technology. Specifically, on average, the proposed solution improves 31% (approximately) of the traditional optimization approaches. But this first evidence is not conclusive, as the distributions overlap. Then, a statistical test must be used to make significant comparisons. Using the Mann-Whitney U statistical test, we found that the proposed solution improves the SLA fulfillment rate (compared to standard optimization approaches) with a significance $p = 0.00075$ (very significant).

4 Conclusions and Future Work

In this paper, we propose a new resource management solution for 6G networks, based on the union of static optimization algorithms and Blockchain-enabled Service Level Agreements, which can be renegotiated dynamically. A transparent Blockchain network allows 6G devices to negotiate their Service Level Agreement with different base stations. Particle swarm optimization algorithms are used to allocate resources and study future potential resource distribution. Results show the Service Level Agreement fulfillment rate increases up to 31%

In future work, simulations in virtual environments, using Software-Defined Networks will be developed.

Acknowledgments. This work is supported by the Ministry of Science, Innovation and Universities through the COGNOS project (PID2019-105484RB-I00).

References

1. Jain, P., Gupta, A., Kumar, N.: A vision towards integrated 6G communication networks: promising technologies, architecture, and use-cases. Phys. Commun. **55**, 101917 (2022)
2. Mandl, P., Pezzei, P., Leitgeb, E.: Comparison of radiation exposure between DVBT2, WLAN, 5G and other sources with respect to law and regulation issues. In: 2020 International Conference on Broadband Communications for Next Generation Networks and Multimedia Applications CoBCom, pp. 1–5. IEEE (2020)
3. Bordel, B., Alcarria, R., Robles, T.: An optimization algorithm for the efficient distribution of resources in 6G verticals. In: Information Systems and Technologies: WorldCIST 2022, vol. 1, pp. 103–114. Springer International Publishing, Cham (2022). https://doi.org/10.1007/978-3-031-04826-5_11
4. Faisal, T., Lucena, J.A.O., Lopez, D.R., Wang, C., Dohler, M.: How to design autonomous service level agreements for 6G. IEEE Commun. Mag. **61**(3), 80–85 (2023)
5. Yang, T., Qin, M., Cheng, N., Xu, W., Zhao, L.: Liquid software-based edge intelligence for future 6G networks. IEEE Netw. **36**(1), 69–75 (2022)
6. Chen, S., Liang, Y.C., Sun, S., Kang, S., Cheng, W., Peng, M.: Vision, requirements, and technology trend of 6G: how to tackle the challenges of system coverage, capacity, user data-rate and movement speed. IEEE Wireless Commun. **27**(2), 218–228 (2020)
7. DEBBABI, F., Rihab, J.M.A.L., CHAARI, L., AGUIAR, R. L., GNICHI, R., TALEB, S.: Overview of AI-based algorithms for network slicing resource management in B5G and 6G. In: 2022 International Wireless Communications and Mobile Computing (IWCMC), pp. 330–335. IEEE (2022)

8. Bhattacharya, P., et al.: A deep-Q learning scheme for secure spectrum allocation and resource management in 6G environment. IEEE Trans. Netw. Serv. Manage. **19**(4), 4989–5005 (2022)
9. Guan, W., Zhang, H., Leung, V.C.: Customized slicing for 6G: enforcing artificial intelligence on resource management. IEEE Netw. **35**(5), 264–271 (2021)
10. Hurtado Sánchez, J.A., Casilimas, K., Caicedo Rendon, O.M.: Deep reinforcement learning for resource management on network slicing: a survey. Sensors **22**(8), 3031 (2022)
11. Sami, H., Otrok, H., Bentahar, J., Mourad, A.: AI-based resource provisioning of IoE services in 6G: a deep reinforcement learning approach. IEEE Trans. Netw. Serv. Manage. **18**(3), 3527–3540 (2021)
12. Mekrache, A., Bradai, A., Moulay, E., Dawaliby, S.: Deep reinforcement learning techniques for vehicular networks: recent advances and future trends towards 6G. Veh. Commun. **33**, 100398 (2022)
13. Prathiba, S.B., Raja, G., Anbalagan, S., Dev, K., Gurumoorthy, S., Sankaran, A.P.: Federated learning empowered computation offloading and resource management in 6G–V2X. IEEE Trans. Netw. Sci. Eng. **9**(5), 3234–3243 (2021)
14. Alsulami, H., Serbaya, S.H., Abualsauod, E.H., Othman, A.M., Rizwan, A., Jalali, A.: A federated deep learning empowered resource management method to optimize 5G and 6G quality of services (QoS). Wireless Commun. Mobile Comput. **2022**, 1352985 (2022)
15. Bordel, B., Alcarria, R., Robles, T.: Interferenceless coexistence of 6G networks and scientific instruments in the K a-band. Expert Syst. e13369 (2023)
16. Bordel, B., Alcarria, R., Robles, T., Sanchez-de-Rivera, D.: Service management in virtualization-based architectures for 5G systems with network slicing. Integr. Comput. Aided Eng. **27**(1), 77–99 (2020)
17. Rasti, M., Taskou, S.K., Tabassum, H., Hossain, E.: Evolution toward 6g multi-band wireless networks: a resource management perspective. IEEE Wireless Commun. **29**(4), 118–125 (2022)
18. Alhashimi, H.F., et al.: A Survey on resource management for 6G heterogeneous networks: current research, future trends, and challenges. Electronics **12**(3), 647 (2023)
19. Fu, S., Wu, B., Wu, S., Fang, F.: Multi-resources management in 6G-oriented terrestrial-satellite network. China Commun. **18**(9), 24–36 (2021)
20. Berardinelli, G., Adeogun, R.: Hybrid radio resource management for 6G subnetwork crowds. IEEE Commun. Mag. **61**(6), 148–154 (2023)
21. Shen, X., Liao, W., Yin, Q.: A novel wireless resource management for the 6G-enabled high-density internet of things. IEEE Wirel. Commun. **29**(1), 32–39 (2022)
22. Zakeri, A., Khalili, A., Javan, M.R., Mokari, N., Jorswieck, E.: Robust energy-efficient resource management, SIC ordering, and beamforming design for MC MISO-NOMA enabled 6G. IEEE Trans. Signal Proc. **69**, 2481–2498 (2021)
23. Long, Q., Chen, Y., Zhang, H., Lei, X.: Software defined 5G and 6G networks: a survey. Mobile Netw. Appl. **27**, 1792–1812 (2019). https://doi.org/10.1007/s11036-019-01397-2
24. Kooshki, F., Rahman, M.A., Mowla, M.M., Armada, A.G., Flizikowski, A.: Efficient Radio Resource Management for Future 6G Mobile Networks: A Cell-less Approach. IEEE Networking Lett. **5**, 95–99 (2023)
25. Xu, H., Klaine, P.V., Onireti, O., Cao, B., Imran, M., Zhang, L.: Blockchain-enabled resource management and sharing for 6G communications. Digit. Commun. Netw. **6**(3), 261–269 (2020)
26. Nyangaresi, V.O., Rodrigues, A.J.: Efficient handover protocol for 5G and beyond networks. Comput. Secur. **113**, 102546 (2022)

Exploring Fatigue in the Workplace: A Data-Driven Approach Using Physiological Signals Captured by a Wristband

Angel Jimenez-Molina[1,2](✉) and Marcelo Riquelme Vicencio[1]

[1] Department of Industrial Engineering Faculty of Physical and Mathematical Sciences, University of Chile, Santiago, Chile
angeljim@uchile.cl, mriquelm@ing.uchile.cl
[2] Engineering Complex Systems Institute, Santiago, Chile

Abstract. Workplace fatigue is a pervasive issue causing accidents, additional costs, and decreased productivity. This paper presents a novel methodology for fatigue identification, aiming to enhance worker wellbeing and business outcomes. The methodology employs wristband captured physiological signals, such as heart rate, and electrodermal activity, to obtain characteristics that serve as objective measures of fatigue. These, when combined with self-reported and observed measures and performance metrics, are used to uncover the fatigue phenomenon using supervised classification methods. The validation of this methodology involved analyzing data from laboratory experiments, where a task mimicking office work was recreated. The Biomonitor wristband was utilized to capture physiological signals in a non-invasive way. The best classification results from the work task experiments was a F1-score of 97.66% using Random Forest. Therefore, the data captured from the Biomonitor wristband and the application of the methodology to identify fatigue were validated. The paper concludes by recommending their use for further research and implementation in industries interested in fatigue management.

Keywords: Fatigue assessment · Psychophysiological signals · Wereable devices

1 Introduction

Occupational fatigue is increasingly recognized worldwide as a problem in modern industry as it is one of the main causes of occupational accidents [13]. The International Labor Organization estimates that every year, more than 313 million workers suffer non-fatal occupational accidents and diseases, which is equivalent to 860,000 victims per day [16], being fatigue a prevalent factor. On the other hand, every day, 6,400 people die due to an occupational accident or occupational disease, and deaths from this cause amount to 2.3 million per year. This leads to estimates that economic losses due to work-related accidents and

J. Bravo and G. Urzáiz (Eds.): UCAmI 2023, LNNS 841, pp. 48–59, 2023.
https://doi.org/10.1007/978-3-031-48590-9_5

illnesses are close to 4% of global GDP, equivalent to 2.8 trillion of US dollars, corresponding to expenses related to lost work time, production stoppages, medical treatment, rehabilitation, and compensation.

Due to the multifactorial causality of fatigue, it is critical and indispensable to have monitoring, classification, and prediction tools to avoid discontinuity in processes and prevent accidents. Mental and physical fatigue occurs in different ways in all work activities, where early detection and good fatigue management can improve people's well-being and productivity indexes. An example is certain office activities in front of a computer, where workers are exposed to monotonous and/or repetitive activities during their workday.

Fatigue is a construct that presents itself as a feeling of tiredness, exhaustion, and difficulty in maintaining task performance [5], related to the cumulative effects of physical and mental stress. In the literature, fatigue is defined in terms of its causes and effects. On the one hand, fatigue is understood as any loss of efficiency in the performance of a task or as the aversion to any kind of effort [6]. On the other hand, fatigue is defined as the psychophysiological result of prolonged periods of physical and cognitive stress [3]. In either case, fatigue produces a decrease in the performance of the sufferer, being a factor in reducing the productivity of a task and the poor quality of its results [2]. The major symptom of mental fatigue is a general feeling of tiredness, feelings of inhibition, and impaired activity. For instance, in the areas of aviation and transportation, mental fatigue is the main cause of serious operating errors [19].

Already in 1908, in [18] Yerkes & Dodson established a model that sought to establish a relationship between stimulus and learning, which, when reinterpreted and applied by different researchers, served to establish a relationship between fatigue and performance [15]. This relationship indicates that the performance of a task increases with physiological or mental activation, but only up to a certain point, and then begins to decrease as fatigue increases.

Fatigue assessment has tended to focus on the use of self-reported instruments and performance measurements, approaches that hinder the monitoring in real time of this construct. A suitable approach is to resort to the nervous system's reaction, through peripheral vital signs, to the stimuli of the work and labor task setting. In fact, physiological signals reflect the Autonomic Nervous System activity that generates unconscious body responses to external stimuli. They have the advantage of being measured in real-time and are not subject to any voluntary response bias of the user; they are considered objective measures to understand the effect of the task on the worker. Therefore, fatigue measurements by means of these are considered objective measures. The literature has shown that physiological signals with the greatest predictive power of fatigue are those captured to obtain heart rate and its variation and electrodermal activity [9].

In this paper, we aim to evaluate to what extent it is possible to assess the fatigue induced by a typical labor task on people by analyzing their psychophysiological signals. To accomplish this goal, we resort to a typing labor task, where study participants must type the information presented in physical invoices in a digital form before a screen. We adopt the self-reported, observed, and performance-based perspectives to measure and validate the fatigue induced to the participants. In addition, we aim to discover the different levels of fatigue users experience during the execution of this task. Whereas participants engage in this typing task, a wristband was used to capture their psychophysiological signals, which are then used to extract meaningful features that feed a classification machine learning algorithm. Thus, we aim to prove the following hypotheses:

- **H1.** A relationship exists between the fatigue levels induced by the task and the features extracted from psychophysiological signals.
- **H2.** It is possible to classify fatigue levels by fusing features extracted from psychophysiological signals.

This paper is organized as follows. Section 2 presents the methods used to perform the study, capture the psychophysiological and behavioral data, and the techniques used for statistical and machine learning analysis. Section 3 presents the results, while the paper is concluded in Sect. 4.

2 Methods

2.1 Participants

The typing task was carried out by 16 volunteer participants, all professionals, including 12 men, aged between 21 and 46 years (mean age = 31.75 years, SD = 8.94 years), with an average of 12.5 years of labor experience and daily computer use. 56% of participants reported engaging in sports on a regular weekly basis, and 75% declared having had a good sleep the previous night (mean sleep hours = 7.32 h, SD = 1.45 h). None of the participants reported cardiovascular diseases or taking medications or recreational drugs that could potentially affect their behavior. The research has the approval of the Ethics Committee of the Faculty of Physical and Mathematical Sciences of the University of Chile. Each participant read and signed an informed consent describing the study procedure, purpose, and right to decline participation at any moment.

2.2 Fatigue-Inducing Task Procedure

Each participant had to type and store the information of a set of physical invoices using a digital form in front of a computer screen, dedicating a total of 110 min to this task. The invoice information included its accounting number, date of the sale, customer data (social number, name, and address), product or service data (name, quantity, unit price), cost accounting number, and payment data (total amount and tax). The experience was carried out in a dedicated, silent room with blackout curtains. Once the participant arrived at the room, she was introduced to the purpose of the study, provided with a description of the typing task, and asked to read and sign the informed consent.

Psychophysiological signals of participants were obtained from the wrist of their non-dominant arm by using the wristband Biomonitor 2.0, developed by

Table 1. The Karolinska Sleepiness Scale (KSS) Questionnaire.

Level	Description
1	Extremely Alert
2	Highly Alert
3	Alert
4	Somewhat Alert
5	Neither Alert nor Drowsy
6	Some sign of Drowsiness
7	Drowsy, no effort to stay awake
8	Drowsy, little effort to stay awake
9	Very Drowsy, high effort to stay awake, struggles with sleep

the WeSST Lab at the University of Chile, about which more information can be found on [7,17]. This wearable device measures a set of biosignals at a high frequency, including photoplethysmography (PPG) at 50 Hz - used to obtain the heart rate (HR) and heart rate variability (HRV) -, electrodermal activity (EDA) at 10 Hz - to derive skin conductance level (SCL) and skin conductance response (SCR) -, skin surface temperature (SKT) at 10 Hz, and three-axis accelerometry and gyroscope data, both measured at a frequency of 50 Hz.

Before performing the typing task, the participant was asked to get used to the device while relaxing music played in the background for ten minutes, after which the experimenter recorded the participant's baseline psychophysiological signals for five minutes. The participant then performed the typing task, whereas the experimenter registered all the incidents and specific user behaviors, and regularly administered the Karolinska Sleepiness Scale (KSS) questionnaire to the participant, with a total of ten applications per experimental session. Further details are provided in the following subsection. Once the session was over, the person was thanked for his or her voluntary participation and asked not to disclose the study procedure to acquaintances to avoid possible bias in future participants.

2.3 Fatigue Measures and Levels Clustering

A three-dimensional measurement was adopted to quantify the fatigue induced by the typing task in the participants through (1) the participant's self-reporting by means of the administration every ten minutes of the KSS questionnaire (see Table 1); (2) the counting and weighting of different, specific gestures and behaviors of the participant by the study monitor in the same periods (see Table 2), and (3) the quantification of the participant's performance in the task by means of the automated identification of typing errors, their frequency, and duration.

Table 2. Participants' Gestures and Behavioral Data to Assess the Observed Fatigue.

Action	Score	Description
Mild sigh	1	Exhale air through mouth and/or nose, slight
Sigh	2	Exhale air through mouth and/or nose, normal
Deep sigh	3	Exhale air through mouth and/or nose, prolonged
Deep breath	3	Inhalation and exhalation through mouth and/or nose, prolonged
Yawn	4	Involuntary action of inhaling air slowly and prolonged, opening the mouth, and then exhaling it also prolonged and sometimes with noise
Stretching	4	Elongation of arms, legs, back and neck
Pause	4	Short period of time where no activity is performed
Chair adjustment	3	Smooth movements in the chair that do not change position
Position adjustment	4	Changes seated position or environment configuration
Sudden movements	5	Movements faster than normal, causing greater auditory impact from the elements used in the task
Asking how much time is left	5	Question asked to know the time, elapsed and/or remaining time of task execution

Self-reported Fatigue. The KSS questionnaire measures the subjective level of self-reported sleepiness during a specific period of time through a Likert-type scale, as shown in Table 1 [1]. The administration of this instrument is triggered by the question "How do you feel?". The KSS questionnaire has been used in studies of shift work, jet lag, and driving, among others, in both men and women.

In [8] Kaida et al. found in a validation study that there is a high correlation between the KSS questionnaire measurements and electroencephalographic signals related to tiredness and fatigue. There is a broad consensus on the advantages of the KSS for its validity, speed, and ease of administration.

Observed Fatigue. Another measure for estimating fatigue is related to the observation of people's behavior through gestures and behaviors that account for the progressive development of fatigue [14], which are weighted according to their relationship with the intensity of the fatigue state, as shown in the Table 2.

Table 3. Features Extracted from the Psychophysiological Signals.

Signal	Feature
EDA	Mean EDA, std EDA, mean SCL, mean SCR, range SCL, max SCR, number of EDA peaks
PPG	Mean HR, std HR, range HR, mean HRV, std HRV, root mean square error (rmss) of HRV, power of the frequency component of the HRV between 0.003 - 0.04 Hz (VLF), power of the frequency component of the HRV between 0.04 - 0.15 Hz (LF), power of the frequency component of the HRV between 0.15 - 0.4 Hz (HF), total power of HRV frequency component (TP), ratio between LF and HF
SKT	Mean SKT, std SKT, range SKT

The KSS scale was positioned above the monitor's height on the wall facing the participant.

Performance Fatigue. Performance measurement methods are frequently used to obtain quantitative information on an individual's ability to sustain attention over time [11]. This is done by identifying behaviors that vary with fatigue levels, such as subtasks execution times, number of errors, and their frequencies. Thus, performance fatigue was quantified through the linear combination of three metrics, which account for (1) effectiveness, measured by the number of errors made by the participant (number of incorrect fields per invoice); (2) efficiency, obtained through the frequency of errors (number of times that at least one error is made per document), and (3) productivity, quantified through the average time T_{m} that the participant requires to type a document and the variance $Var(T_{\mathrm{m}})$ of that time. The first two metrics are used to determine the error indicator I_{e}, given by $I_{\mathrm{e}} = pF_{\mathrm{e}} + (1-p)N_{\mathrm{e}}$, where N_{e} represents the number of errors, F_{e} their frequency, and p is an importance weight for F_{e}. In this way, performance fatigue is determined by $F_{\mathrm{p}} = \alpha I_{\mathrm{e}} + \beta T_{\mathrm{m}} + \gamma Var(T_{\mathrm{m}})$.

Each dimension used to measure the fatigue induced in the participants by the typing task was standardized by applying a z-scoring to their values. This three-dimensional measurement of fatigue resulted in the three-dimensional labeling of a set of ten-minute segments. These segments were clustered to identify the different levels of fatigue induced by the typing task. The optimal number of clusters was calculated with the criterion of the Calinski Harabasz index, which is based on maximizing the ratio of the between-cluster variance and the total intra-cluster variation [4]. The clustering was performed using the k-means clustering method. Next, to determine the correspondence between each group and each level of fatigue, the sum of each dimension of the cluster centroid is calculated and ordered from lowest to highest, where the minimum corresponds to the lowest level and the maximum to the highest. This is valid because the

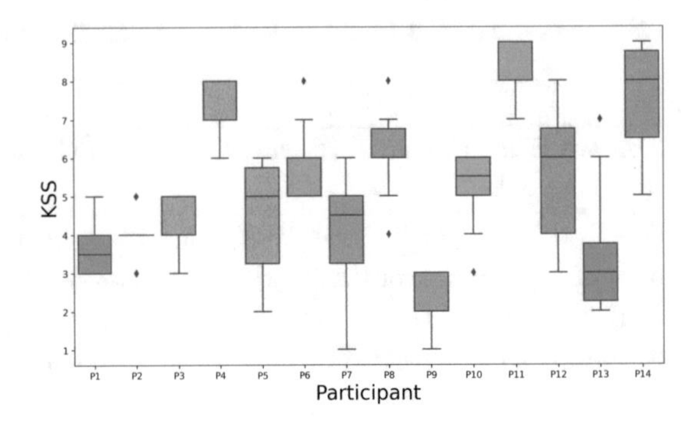

Fig. 1. Box plot of Participants' KSS Responses in each Ten-minute Segment of the Typing Task.

three measures of fatigue, self-reported, observed, and performance, are directly proportional to total fatigue [2].

To visualize the progression of fatigue levels, we performed a temporal mapping of the total fatigue of each participant. In addition, we averaged the participants' total fatigue in each segment to show the trend of induced fatigue in the entire group of users.

2.4 Features Extraction and Fatigue Classification

According to the evidence reported by [10,12] we adopted a strategy of partitioning and sliding the signals into five-minute sections with ten seconds of sliding. We refer to these sections as fatigue assessment windows, which account for the progressive nature of fatigue over time. These windows correspond to the instances used for training and testing the fatigue classification models.
Table 3 shows the 21 psychophysiological features extracted from the preprocessed signals at each fatigue assessment window that, according to the literature, are meaningful to classify fatigue in diverse settings. In addition, we include the averages of the accelerometry and gyroscope signals across three axes in this set of features.

To test hypothesis 1 about the existence of a relationship between the fatigue levels induced by the typing task and the features extracted from psychophysiological signals, we use the Kruskal-Wallis non-parametric statistical test by comparing the medians of these features across the fatigue levels. On the other hand, we apply the Boruta algorithm to reduce the dimensionality of the set of extracted psychophysiological features, whereas the classification task (hypothesis 2) is performed by using Support Vector Machine (SVM), Feedforward Neural Network (FFN), and Random Forest (RF) machine learning algorithms. To avoid overfitting, the validation is conducted using a three-times repeated 10-fold cross-validation and using F1-score, accuracy, precision, recall, and kappa as the evaluation metrics. The kappa statistic contrasts the accuracy that could be obtained with any random classifier applied to imbalanced classes against the accuracy achieved with the machine learning algorithm under evaluation. Therefore, this statistic is appropriate to assess whether a metric like accuracy is distorted due to the risk for biased classification arising from class imbalance.

3 Results

Figure 1 shows the KSS questionnaire bloxplots for the entire group. These values represent the participants' self-perception of the induced fatigue. A Cronbach-alpha statistic equal to 0.976 confirms the internal consistency and reliability of the answers given by the participants.

The criterion of the Calisnski Harabasz index yielded three clusters as the optimal number of groups for clustering the set of ten-minute, three-dimensional labeled segments. Figure 2 shows the result of applying the K-means clustering algorithm to these segments and the centroid of each of the three groups. By analyzing the values of the centroids, we associate them and their segments to three classes: without fatigue, low fatigue, and high fatigue. The progression of these fatigue levels for some of the participants is shown in Fig. 3.

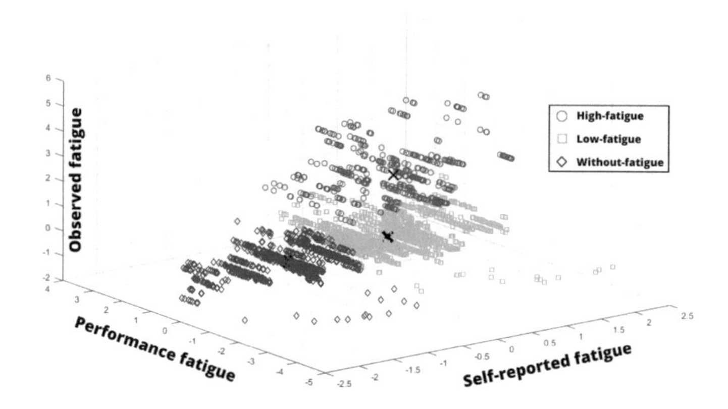

Fig. 2. Three-dimensional Fatigue Labels Clustering based on Self-reported, Performance, and Observed Fatigue.

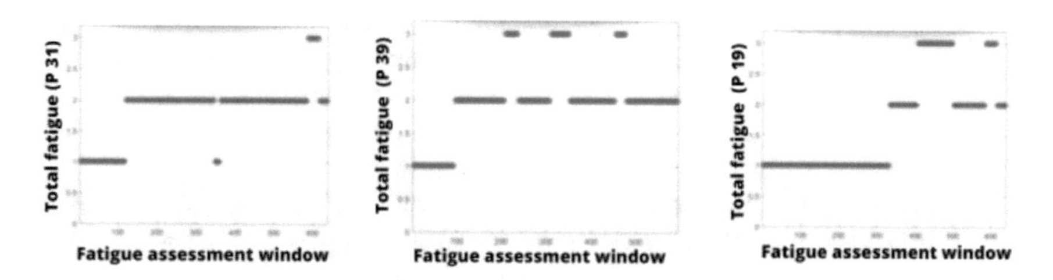

Fig. 3. Example of Fatigue Labels and their Temporal Evolution for Participants 31, 39, and 19 (1: without fatigue; 2: low fatigue; 3: high fatigue).

Table 4. Comparison of the Median of the Psychophysiological Features Across the Three Fatigue Levels Using the Kruskal-Wallis Test.

Feature	Without fatigue (median)	Low fatigue (median)	High fatigue (median)	Tendency	Significance level
Mean HR	−0.017	0.068	0.083	Upward	***
Std HR	0.816	0.852	0.847	-	***
Range HR	4.021	4.541	3.980	-	***
Mean HRV	0.022	0.022	−0.032	Downward	***
Std HRV	0.936	0.969	0.954	-	***
Rmssd HRV	0.006	0.007	0.006	-	***
VLF	0.535	0.512	0.506	-	”
LF	0.531	0.508	0.505	Downward	**
HF	0.526	0.501	0.504	-	***
TP	0.470	0.439	0.449	-	***
Ratio LF/HF	1.029	1.010	1.043	-	”
Mean EDA	−0.229	−0.241	−0.265	Downward	***
Std EDA	0.075	0.104	0.150	Upward	***
Mean SCL	−0.102	−0.085	−0.200	-	***
Mean SCR	−0.175	−0.017	−0.005	Upward	***
Range SCL	0.237	0.306	0.392	Upward	***
Max SCR	0.285	0.519	1.800	Upward	***
Npeak SCR	1.000	3.000	3.000	Upward	***
Mean SKT	0.005	0.281	0.029	-	***
Std SKT	0.096	0.084	0.101	-	***
Range SKT	0.370	0.299	0.353	-	***

Note: *** $p < 0.001$; ** $p < 0.01$; ” $p > 0.1$.

This behavior of fatigue is coherent with the assumption that there does not exist fatigue at the beginning of the typing task and also provides the insight that after reaching a high level of fatigue, participants show periods of recovering to low fatigue. Nevertheless, it is rare for participants to experience a complete recovery, returning to a state without fatigue, which confirms that the typing task is valid as a means to induce fatigue and therefore use it to study its psychophysiological correlates. Moreover, on average, for the entire group of participants, fatigue increased during the 10-minute segments.

The partition and sliding strategy into five minutes and ten seconds sliding yielded a total of 10,128 fatigue assessment windows.

Table 4 shows that several psychophysiological features have statistically significant differences in their median values across the three levels of fatigue ($p < 0.001$), which implies that these psychophysiological features are not distributed the same across these levels. This result proves hypothesis 1 about a relationship between the fatigue levels induced by the typing task and the features extracted from psychophysiological signals.

Table 5. Fatigue Level Classification Performance Metrics of Different Machine Learning Algorithms for the Typing Task.

Model	Accuracy (%)	Precision (%)	Recall (%)	F1-score (%)	Kappa (%)
SVM	83.11	87.78	67.27	70.40	66.44
FFN	75.21	77.53	80.89	78.12	48.35
RF	98.45	97.98	97.36	97.66	97.20

Also, these results identify, on the one hand, an increasing pattern between the levels of fatigue for the mean of the HR, the standard deviation of the EDA, the mean of the SCR, the range of the SCL, the maximum of the SCR, and the number of peaks of the SCR. On the other hand, a decreasing pattern is observed for the mean of the HRV and the mean of the EDA. In addition, a set of psychophysiological features that do not show statistical significance are the VLF, HF, the ratio LF/HF, and the mean SCL. This last result is worth to be analyzed in future research to assess to what extent the characteristics of the labor task is determinant in the reaction of the nervous system and, therefore, of the peripheral vital signs since HF, LF, and VLF have been related to drowsiness in some studies, especially in driving settings.

Table 5 shows the classification results (in the testing folds of the 10-folds partition) of the three classes (without fatigue, low fatigue, and high fatigue) using the three classifiers (SVM, FFN and RF) and the features extracted from the psychophysiological signals at each fatigue assessment window and selected by the Boruta algorithm. F1-score is the most suitable evaluation metric in a setting of unbalanced labeling. The worst result for this metric is obtained by the SVM machine learning algorithm, which reaches an F1-score of 70.4%. A better result is obtained by the FFN (78.12%) but with poor Kappa statistics (48.35%), which indicates that this model instance is impacted by the imbalance of classes, which produces a bias in the classification. The results improve with the RF machine learning algorithm, where both the F1-score and the Kappa statistics have high values, 97.66%, and 97.2%, respectively, which proves hypothesis 2 about the possibility of classifying fatigue levels by fusing features extracted from psychophysiological signals.

4 Conclusion

This paper presents a data-driven approach, based on the analysis of psychophysiological signals captured by a wristband, to assess the fatigue induced by labor tasks in people. A typing task where a set of professional users had to type the information of a physical invoice in a digital form in front of the screen of a computer during 110 min, whereas their psychophysiological signals were captured, was used as a typical task in an office work setting. The fatigue was measured every ten-minute segment using three dimensions: self-reported fatigue with the KSS questionnaire, observed fatigue through indicative gestures and behaviors, and performance fatigue based on errors, their frequency, and subtasks' time completion. A linear combination of these dimensions yielded a total fatigue per segment and participant, which average value for the entire group shows a growing tendency over time, confirming the suitability of this typing task to induce fatigue. A clustering analysis revealed that participants experienced three levels of fatigue with this task, beginning without fatigue and then gradually increasing to low and high fatigue, interleaving with moments of recovery. Adopting a strategy of five minutes partitioning with slicing of ten seconds to generate a set of fatigue assessment windows, different statistical and frequency domain features were extracted from the psychophysiological signals. By using the labeling created from the three-dimensional approach, several features were shown to be statistically significant in differentiating among the three levels of fatigue, such as the mean of the HR, the standard deviation of the EDA, the mean of the SCR, the range of the SCL, the maximum of the SCR, and the number of peaks of the SCR, all of which shown an increasing pattern among the levels. Other features, such as the mean of the HRV, showed a decreasing pattern, which is coherent with what has been found in the literature. Finally, by resorting to classic machine learning algorithms, it was proved that it is possible to classify the three found levels of fatigue based on the selected psychophysiological features, reaching an F1-score of 97.66% and a kappa statistics of 97.2% with the RF model. These results are meaningful in informing fatigue risk systems with insightful behavioral data that can be used to improve managerial decisions in the workplace, such as staff shifting or resting protocols, among others, which can have a positive impact on the well-being of workers and productivity. In our future work, we plan to replicate this approach on data we have captured in different work settings, such as car and bus driving and physical labor tasks. We are especially interested in differentiating the fixed and individual effects such labor tasks can have on heterogeneous workers from a causal point of view. Also, with this empirical experience, we plan to devise a general framework for fatigue assessment based on psychophysiological indicators.

Acknowledgements. This research was partially funded by ANID, FONDECYT 11130252; ANID, PIA/BASAL AFB180003; Proyecto Superintendencia de Seguridad Social (SUSESO)IST: ID83, IST201783; IMA+ Innovacion en Manufactura Avanzada. Research assistantship from Jorge Gaete, Cristian Retamal, Ignacio Vargas, and Francisco Díaz is greatly acknowledged.

References

1. Åkerstedt, T., Gillberg, M.: Subjective and objective sleepiness in the active individual. Int. J. Neurosci. **52**(1–2), 29–37 (1990)
2. Aryal, A., Ghahramani, A., Becerik-Gerber, B.: Monitoring fatigue in construction workers using physiological measurements. Autom. Constr. **82**, 154–165 (2017)
3. Boksem, M.A., Tops, M.: Mental fatigue: costs and benefits. Brain Res. Rev. **59**(1), 125–139 (2008)
4. Caliński, T., Harabasz, J.: A dendrite method for cluster analysis. Commun. Stat.-Theory Methods **3**(1), 1–27 (1974)

5. Council, N.S.: www.nsc.org, [Acceso revisado, 16-Mayo-2019]
6. Grandjean, E.: Fatigue in industry. Occup. Environ. Med. **36**(3), 175–186 (1979)
7. Jimenez-Molina, A., Diaz-Guerra, F., Retamal, C., Guevara, C.A.: Towards psychophysiological markers for affect-aware vehicles. In: Bravo, J., Ochoa, S., Favela, J. (eds.) Proceedings of the International Conference on Ubiquitous Computing & Ambient Intelligence (UCAmI 2022), pp. 571–582. Springer, Cham (2023). https://doi.org/10.1007/978-3-031-21333-5_58
8. Validation of the karolinska sleepiness scale against performance and EEG variables. Clin. Neurophysiol. **117**(7), 1574–1581 (2006)
9. Mohanavelu, K., Lamshe, R., Poonguzhali, S., Adalarasu, K., Jagannath, M.: Assessment of human fatigue during physical performance using physiological signals: a review. Biomed. Pharmacol. J. **10**(4), 1887–1896 (2017)
10. Patel, M., Lal, S.K., Kavanagh, D., Rossiter, P.: Applying neural network analysis on heart rate variability data to assess driver fatigue. Expert Syst. Appl. **38**(6), 7235–7242 (2011)
11. Riccio, C.A., Reynolds, C.R., Lowe, P., Moore, J.J.: The continuous performance test: a window on the neural substrates for attention? Arch. Clin. Neuropsychol. **17**(3), 235–272 (2002)
12. von Rosenberg, W., Chanwimalueang, T., Adjei, T., Jaffer, U., Goverdovsky, V., Mandic, D.P.: Resolving ambiguities in the LF/HF ratio: LF-HF scatter plots for the categorization of mental and physical stress from hrv. Front. Physiol. **8**, 360 (2017)
13. Sadeghniiat-Haghighi, K., Yazdi, Z.: Fatigue management in the workplace. Ind. Psychiatry J. **24**(1), 12 (2015)
14. Sahayadhas, A., Sundaraj, K., Murugappan, M.: Detecting driver drowsiness based on sensors: a review. Sensors **12**(12), 16937–16953 (2012)
15. Teigen, K.H.: Yerkes-dodson: a law for all seasons. Theory & Psychology **4**(4), 525–547 (1994)
16. del Trabajo, O.I.: Construir una cultura de prevención en materia de seguridad y salud en el trabajo. https://www.ilo.org/global/about-the-ilo/how-the-ilo-works/ilo-director-general/statements-and-speeches/WCMS_364085/lang--es/index.htm (2015). Accessed 15 July 2023
17. WeSST Lab, W.S., Lab, S.T.: Biomonitor v2.0 wristband. https://isci.cl/biomonitor/ (2019). Accessed 15 July 2023
18. Yerkes, R.M., Dodson, J.D.: The relation of strength of stimulus to rapidity of habit-formation. J. Comp. Neurol. Psychol. **18**(5), 459–482 (1908)
19. Zhao, C., Zheng, C., Zhao, M., Liu, J.: Physiological assessment of driving mental fatigue using wavelet packet energy and random forests. Am. J. Biomed. Sci. **2**(3), 262–274 (2010)

Federated Learning for Industry 5.0: A State-of-the-Art Review

Tamai Ramírez[1(✉)], Eduardo Calabuig-Barbero[2], Higinio Mora[1], Francisco A. Pujol[1], and Sandra Amador[1]

[1] Department of Computer Technology and Computation, University of Alicante, Alicante, Spain
tamai.ramirez@ua.es
[2] Footwear Technology Center, 03600 Elda, Spain
https://web.ua.es/es/dtic/

Abstract. Federated Learning (FL) and Industry 5.0's convergence holds significant promise for changing smart systems. FL, a distributed machine learning method, allows for collaborative model training while protecting data privacy and security. The purpose of this review is to establish the current state-of-the-art on FL and its possible applications in Industry 5.0. Key accomplishments, problems, and future research paths are outlined through detailed literature review. FL has various advantages for Industry 5.0, including increased model accuracy, reduced communication overhead, and improved data privacy. Due to its decentralized nature, it allows for local model training on edge devices, facilitating real-time decision-making and lowering latency. Furthermore, the research investigates FL's potential influence on smart cities by enabling dispersed learning across urban domains without the need for data exchange and how could improve the performance in manufacturing tasks and its role in enhancing Human-Robot Interaction (HRI), resulting in safer and more efficient industrial operations. This gives unparalleled opportunity for developing intelligent systems that address the challenges brought by Industry 5.0 paradigm. The provided state-of-the-art review aims to inform researchers and practitioners about the current advancements in FL and serves as a foundation for future studies to harness the full potential of this novel paradigm for a wide range of industries and applications.

Keywords: Federated Learning · Industry 5.0 · Machine Learning · Internet of Things

1 Introduction

The industrial environment is primed for a revolutionary change with the arrival of Industry 5.0. This paradigm stresses the integration of human knowledge and skills with modern technology, encouraging seamless human-machine collaboration. Industry 5.0 intends to leverage the potential of automation, artificial

J. Bravo and G. Urzáiz (Eds.): UCAmI 2023, LNNS 841, pp. 60–66, 2023.
https://doi.org/10.1007/978-3-031-48590-9_6

intelligence, robotics, and the Internet of Things (IoT) to empower and augment human workers rather than replace them, building on the foundations set by Industry 4.0 [8,24].

While previous technical advances have resulted in more automation in industry, individuals' particular skills, resourcefulness, and problem-solving abilities have frequently been disregarded. In response, Industry 5.0 promotes human involvement in industrial context decision-making, problem-solving, and innovation. One of the primary goals of Industry 5.0 is to create a collaborative environment in which humans and machines complement each other's strengths. Humans are involved in complex problem solving, critical thinking, creativity, and social relations, whereas machines are involved in repetitious and physically demanding jobs. This collaboration not only boosts productivity and efficiency, but it also encourages adaptation and creativity in response to changing market demands [1].

In addition, Industry 5.0 anticipates a trend toward individualized and customized production. Manufacturing processes can be adjusted to particular client needs by combining current technology and exploiting data-driven insights, enabling mass production of highly personalized products. This level of personalization blends human skill in creating and optimizing manufacturing systems with machine precision, speed, and automation [6,12].

Federated Learning has emerged as a compelling method associated with Industry 5.0 concepts in this changing context. FL allows for collaborative machine learning while maintaining data privacy and security. It enables decentralized machine learning model training over several devices, with data saved locally to guarantee privacy. FL contributes to Industry 5.0 aims by actively incorporating human workers in the model training process and exploiting the collective intelligence of distributed devices [10,20].

As FL develops pace across multiple sectors, a cutting-edge examination of its applications and successes in the context of Industry 5.0 becomes essential. This type of review is critical for researchers, practitioners, and policymakers to understand the current state of FL in Industry 5.0, to identify important problems and outstanding topics, and to provide ideas for effective implementation. This study intends to undertake a thorough review of existing literature in order to throw light on the benefits, challenges, and future directions for the integration of FL in Industry 5.0, thereby contributing to the advancement of this collaborative learning paradigm.

2 State of the Art

Federated Learning has emerged as a blossoming and fast maturing field in recent years, garnering significant attention. Given its dynamic nature, it is critical to focus attention on understanding the domain's problems and unresolved topics. As a result, this section intends to conduct a comprehensive overview of the most recent and prominent contributions in FL, revealing present advances and illuminating future research paths.

First and foremost, it is critical to describe the key characteristics and assumptions that form a typical FL setup. These characteristics are important in determining the landscape of FL research and practice. The following are the essential features and assumptions:

(1) Non-IID Data: Because training data in FL is created by a variety of devices and users, the assumption of independent and identically distributed (IID) training data is erroneous. The data distribution among devices is heterogeneous, providing a considerable problem in properly aggregating model updates [15,29]. (2) Optimization Algorithms: FL primarily relies on iterative learning procedures including a sequence of client-server interactions known as training rounds. These rounds are designed to improve a global model by combining local model updates from participating devices. The selection and design of efficient optimization algorithms is critical to attaining efficient and accurate model convergence [7]. (3) Security and Privacy: Ensuring data privacy and security is a fundamental concept of FL. FL avoids the need to communicate raw data by design, instead focusing on exchanging only model weights or changes. This decentralized architecture, in which local models remain on individual devices, protects sensitive data privacy, making FL an appealing alternative for privacy-conscious applications [18,26]. (4) Heterogeneous Devices: FL uses a decentralized learning paradigm in which several edge devices contribute to the training process. However, dynamic computation and communication conditions can have an impact on the performance of these devices. The heterogeneity of devices makes it difficult to coordinate their participation, accommodate variances in computational capacity, and manage communication restrictions [23].

These four vital features and assumptions underpin the essence of FL and significantly shape the research and development efforts in this field. Understanding and addressing the implications of these features are crucial for advancing the state of the art in FL and overcoming the associated challenges [3].

Federated Learning is useful when local devices do not have enough data or labeled instances. FL provides model aggregation while avoiding the sharing of raw data, tackles data scarcity, and assures secure model sharing. Communication costs, privacy hazards during training, and device variability, on the other hand, offer obstacles. Advancing FL's efficacy and enabling collaborative, privacy-preserving, and efficient decentralized machine learning requires optimizing communication, inventing privacy-enhancing approaches, and resolving device heterogeneity [7].

Because Federated Learning is distributed, intelligent Internet of Things applications and systems capable of accommodating FL algorithms are required. As a result, subsequent studies and academics have thoroughly investigated the remaining obstacles connected with deploying FL over IoT networks. As previously stated, allowing FL in systems with a large number of heterogeneous devices running under strict wireless resource constraints is a significant problem in this domain. A proposed method entails using sparsification techniques to construct a criterion for determining which subset of smaller devices should communicate their data to the edge/cloud server. However, determining the proper criterion, such as picking devices with gradients above a certain threshold, poses a significant issue. Nonetheless, establishing the best threshold value in a huge and heterogeneous system is a daunting computing challenge [11,25,27].

Another aspect mentioned is security that also needs to be ensure in IoT systems. Because of its decentralized and unchangeable nature, Blockchain technology is becoming increasingly popular in IoT applications. It improves data integrity, secrecy, and trust among Internet of Things devices, allowing for secure and auditable transactions. The distributed consensus mechanism of Blockchain provides dependable and tamper-resistant interactions, making it a promising alternative for improving the security and privacy of IoT devices [2,4,16]. The use of Blockchain technology in conjunction with Federated Learning in IoT systems improves security and privacy. By exploiting its decentralized and immutable nature, blockchain secures the integrity and confidentiality of IoT data during collaborative model training. It functions as a secure ledger for verifiable and auditable transactions, allowing for secure model aggregation while preventing manipulation or illegal access. This integration promotes confidence among participating devices, resulting in a secure and private learning environment in FL-based IoT systems [9,19,22,30].

The Industrial Internet of Things (IIoT) is a game-changing technology that connects industrial gadgets, machines, and systems. It offers real-time data interchange, process optimization, and increased operational efficiency. The IoT provides organizations with automation, predictive maintenance, and data-driven decision-making, transforming industries and fostering a smarter, more linked industrial ecosystem [21]. The authors of [5] give an in-depth examination of the obstacles and future perspectives for enabling Federated Learning in Industrial Internet of Things (IIoT) systems. While some of the issues are similar to those addressed previously, such as optimization techniques, data privacy, and communication overhead, the authors also identify some new ones. These challenges include: (1) anomaly detection, which focuses on understanding how failures in edge devices can impact overall production processes; (2) inference attack, in which malicious entities can deduce sensitive information from the results of permitted queries; and (3) automation of industrial processes, emphasizing the importance of reducing human labor as a critical step toward achieving complete automation in smart industries. The insights provided by the authors shed light on the unique challenges specific to FL in IIoT and offer valuable directions for future research in this domain.

3 Challenges and Open Issues

Recent advances in the field of Federated Learning have highlighted the orchestration of devices with cloud infrastructure to improve overall system efficiency when Federated Learning is used. Certain recent research attempts have pushed for the implementation of Lyapunov optimization theory as part of these developments [28]. Furthermore, in light of recent advancements within the domain of Natural Language Processing, Federated Learning is being harnessed to augment efficiency in tasks like text recognition. Owing to its inherent heterogeneity, Federated Learning contributes to the enrichment of data derived from a multitude of devices and users [14].

As the growth and use of Federated Learning technology spreads to new systems, new issues develop that involve the academic community. These include: (1) the feasibility of FL in mitigating cyberattacks and protecting data privacy within IoT systems [17], (2) the application of FL algorithms in preventing accidents through the detection of changes in road conditions and the acquisition of adaptive responses in autonomous vehicles, and (3) the potential of FL to provide effective solutions for addressing battery consumption issues in IoT devices executing Machine Learning models [3].

The non-IID (Non-Identically and Independently distributed) character of data distribution poses a significant issue in the field of Federated Learning, deviating from the traditional assumption of homogeneously dispersed data in centralized machine learning frameworks. The presence of this variability across dispersed datasets might limit model generalization, slow convergence rates, raise privacy concerns, and necessitate the adaption of underlying algorithms. Ongoing research efforts are intensively examining strategies to alleviate these issues and improve the durability of Federated Learning in situations with diverse data distribution patterns. Despite the fact that several recent studies have been dedicated to resolving this issue, it remains a daunting task in the present research scene [13,15,29].

4 Conclusions and Future Work

This paper focuses on the future integration of Industry 5.0 with Federated Learning, with an emphasis on human-machine collaboration and decentralized machine learning for personalized production and data privacy protection. Future research should focus on optimizing communication and decreasing latency, building robust privacy-preserving algorithms, studying federated transfer learning and lifetime federated learning, and addressing device heterogeneity to fully realize FL's potential in Industry 5.0. Bridging the knowledge gap between humans and machines will result in intelligent, efficient, and human-centric industrial systems that will promote sustainable practices and drive innovation in manufacturing and elsewhere. This comprehensive examination is an excellent resource for scholars and practitioners, inspiring breakthroughs that will influence the transformative future of Industry 5.0 and FL, bringing us closer to a vibrant and inclusive industrial environment.

Acknowledgments. This work was supported by the Spanish Research Agency (AEI) under project HPC4Industry PID2020-120213RB-I00 (DOI: 10.13039/501100011033).

References

1. Adel, A.: Future of industry 5.0 in society: human-centric solutions, challenges and prospectiveăresearch areas. J. Cloud Comput. **11**(1), 40 (2022). https://doi.org/10.1186/s13677-022-00314-5
2. Ayub Khan, A., Laghari, A.A., Shaikh, Z.A., Dacko-Pikiewicz, Z., Kot, S.: Internet of things (IoT) security with blockchain technology: a state-of-the-art review. IEEE Access **10**, 122679–122695 (2022). https://doi.org/10.1109/ACCESS.2022.3223370
3. Banabilah, S., Aloqaily, M., Alsayed, E., Malik, N., Jararweh, Y.: Federated learning review: fundamentals, enabling technologies, and future applications. Inf. Process. Manag. **59**(6), 103061 (2022). https://doi.org/10.1016/j.ipm.2022.103061
4. Barański, S., Szymański, J., Mora, H.: Anonymous provision of privacy-sensitive services using blockchain and decentralised storage. Res. Square (2023)
5. Boobalan, P., et al.: Fusion of federated learning and industrial internet of things: a survey. Comput. Netw. **212**, 109048 (2022). https://doi.org/10.1016/j.comnet.2022.109048
6. Coelho, P., Bessa, C., Landeck, J., Silva, C.: Industry 5.0: the arising of a concept. Procedia Comput. Sci. **217**, 1137–1144 (2023). https://doi.org/10.1016/j.procs.2022.12.312
7. Elouali, A., Mora Mora, H., Mora-Gimeno, F.J.: Data transmission reduction formalization for cloud offloading-based IoT systems. J. Cloud Comput. **12**(1), 1–12 (2023). https://doi.org/10.1186/s13677-023-00424-8
8. Golovianko, M., Terziyan, V., Branytskyi, V., Malyk, D.: Industry 4.0 vs industry 5.0: co-existence, transition, or a hybrid. Procedia Comput. Sci. **217**, 102–113 (2023). https://doi.org/10.1016/j.procs.2022.12.206
9. Issa, W., Moustafa, N., Turnbull, B., Sohrabi, N., Tari, Z.: Blockchain-based federated learning for securing internet of things: a comprehensive survey. ACM Comput. Surv. **55**(9), 1–43 (2023). https://doi.org/10.1145/3560816
10. Khan, F., Kumar, R.L., Abidi, M.H., Kadry, S., Alkhalefah, H., Aboudaif, M.K.: Federated split learning model for industry 5.0: a data poisoning defense for edge computing. Electronics **11**(15), 2393 (2022). https://doi.org/10.3390/electronics11152393
11. Khan, L.U., Saad, W., Han, Z., Hossain, E., Hong, C.S.: Federated learning for internet of things: recent advances, taxonomy, and open challenges. IEEE Commun. Surv. Tutor. **23**(3), 1759–1799 (2021). https://doi.org/10.1109/COMST.2021.3090430
12. Leng, J., et al.: Industry 5.0: prospect and retrospect. J. Manuf. Syst. **65**, 279–295 (2022). https://doi.org/10.1016/j.jmsy.2022.09.017
13. Li, Q., Diao, Y., Chen, Q., He, B.: Federated learning on non-iid data silos: an experimental study. In: 2022 IEEE 38th International Conference on Data Engineering (ICDE), pp. 965–978 (2022). https://doi.org/10.1109/ICDE53745.2022.00077
14. Liu, F., Wu, X., Ge, S., Fan, W., Zou, Y.: Federated learning for vision-and-language grounding problems. In: Proceedings of the AAAI Conference on Artificial Intelligence, vol. 34, pp. 11572–11579 (2020). https://doi.org/10.1609/aaai.v34i07.6824
15. Ma, X., Zhu, J., Lin, Z., Chen, S., Qin, Y.: A state-of-the-art survey on solving non-iid data in federated learning. Fut. Gener. Comput. Syst. **135**, 244–258 (2022). https://doi.org/10.1016/j.future.2022.05.003

16. Mora, H., Mendoza-Tello, J.C., Varela-Guzmáin, E.G., Szymanski, J.: Blockchain technologies to address smart city and society challenges. Comput. Human Behav. **122**, 106854 (2021). https://doi.org/10.1016/j.chb.2021.106854
17. Mora, H., Pujol, F.A., Ramírez, T., Jimeno-Morenilla, A., Szymanski, J.: Network-assisted processing of advanced iot applications: challenges and proof-of-concept application. Cluster Comput. 1–17 (2023). https://doi.org/10.1007/s10586-023-04050-6
18. Mothukuri, V., Parizi, R.M., Pouriyeh, S., Huang, Y., Dehghantanha, A., Srivastava, G.: A survey on security and privacy of federated learning. Fut. Gener. Comput. Syst. **115**, 619–640 (2021). https://doi.org/10.1016/j.future.2020.10.007
19. Prokop, K., Połap, D., Srivastava, G., Lin, J.C.W.: Blockchain-based federated learning with checksums to increase security in internet of things solutions. J. Ambient. Intell. Humaniz. Comput. **14**(5), 4685–4694 (2023). https://doi.org/10.1007/s12652-022-04372-0
20. Singh, S.K., Yang, L.T., Park, J.H.: Fusionfedblock: fusion of blockchain and federated learning to preserve privacy in industry 5.0. Inf. Fusion **90**, 233–240 (2023). https://doi.org/10.1016/j.inffus.2022.09.027
21. Sisinni, E., Saifullah, A., Han, S., Jennehag, U., Gidlund, M.: Industrial internet of things: challenges, opportunities, and directions. IEEE Trans. Ind. Inf. **14**(11), 4724–4734 (2018). https://doi.org/10.1109/TII.2018.2852491
22. Visvizi, A., Mora, H., Varela-Guzman, E.G.: The case of rwallet: a blockchain-based tool to navigate some challenges related to irregular migration. Comput. Hum. Behav. **139**, 107548 (2023). https://doi.org/10.1016/j.chb.2022.107548
23. Xu, C., Qu, Y., Xiang, Y., Gao, L.: Asynchronous federated learning on heterogeneous devices: a survey. arXiv preprint arXiv:2109.04269 (2021)
24. Xu, X., Lu, Y., Vogel-Heuser, B., Wang, L.: Industry 4.0 and industry 5.0 inception, conception and perception. J. Manuf. Syst. **61**, 530–535 (2021). https://doi.org/10.1016/j.jmsy.2021.10.006
25. Yang, Z., Chen, M., Wong, K.K., Poor, H.V., Cui, S.: Federated learning for 6G: applications, challenges, and opportunities. Engineering **8**, 33–41 (2022). https://doi.org/10.1016/j.eng.2021.12.002
26. Zhang, K., Song, X., Zhang, C., Yu, S.: Challenges and future directions of secure federated learning: a survey. Front. Comp. Sci. **16**, 1–8 (2022). https://doi.org/10.1007/s11704-021-0598-z
27. Zhang, T., Gao, L., He, C., Zhang, M., Krishnamachari, B., Avestimehr, A.S.: Federated learning for the internet of things: applications, challenges, and opportunities. IEEE Internet Things Maga. **5**(1), 24–29 (2022). https://doi.org/10.1109/IOTM.004.2100182
28. Zhou, Z., Yang, S., Pu, L., Yu, S.: CEFL: online admission control, data scheduling, and accuracy tuning for cost-efficient federated learning across edge nodes. IEEE Internet Things J. **7**(10), 9341–9356 (2020)
29. Zhu, H., Xu, J., Liu, S., Jin, Y.: Federated learning on non-iid data: a survey. Neurocomputing **465**, 371–390 (2021). https://doi.org/10.1016/j.neucom.2021.07.098
30. Zhu, J., Cao, J., Saxena, D., Jiang, S., Ferradi, H.: Blockchain-empowered federated learning: challenges, solutions, and future directions. ACM Comput. Surv. **55**(11), 1–31 (2023). https://doi.org/10.1145/3570953

Discriminating Deceptive Energy Generation of Photovoltaic Systems by Deep Learning and Adversarial Networks

Aurora Polo-Rodriguez[1,4], Guillermo Almonacid-Olleros[2], Gabino Almonacid[2], Chris Nugent[3], and Javier Medina-Quero[4](✉)

[1] Department of Computer Science, University of Jaén, 23071 Jaén, Spain
apolo@ujaen.es
[2] Department of Electronic Engineering. Campus Las Lagunillas, 23071 Jaén, Spain
[3] School of Computing, Ulster University. Northern Ireland, Ireland, UK
[4] Department of Computer Engineering, Automation and Robotics, University of Granada, 18071 Granada, Spain
javiermq@ugr.es

Abstract. In this work, we evaluate the capabilities of Deep Learning and Adversarial Networks to nowcast and discriminate the output power generation in photovoltaic systems. From a baseline spatiotemporal model of Convolutional Neural Networks and Long Short-Term Memories, we develop a discriminator and generator based on Conditional Generative Adversarial Networks. The adversarial network develops the estimation and discrimination of erroneous output power generation. Two real-world datasets are evaluated with encouraging results that are straightforwardly related to maintenance deployment applications in photovoltaic systems.

Keywords: Energy generation · Photovoltaic Systems · Adversarial Networks

1 Introduction

Photovoltaic (PV) power generation has seen remarkable success and maturity, with projections indicating a substantial global increase in renewable power capacity over the next five years [11]. Consequently, the management of operation and maintenance in the solar photovoltaic (PV) industry has become a critical research focus [8]. In this context, data plays a pivotal role, serving as a valuable asset for modelling the system's standard behaviour and closely monitoring its performance against the predicted output derived from the model. This comprehensive monitoring, encompassing all factors that may impact performance, enables early detection of damages and faults [5].

Within the field of energy systems, deep learning (DL) has emerged as a highly promising and innovative approach, particularly in the areas of photovoltaic (PV) power nowcasting or forecasting (PVPNF) [15] and fault detection

J. Bravo and G. Urzáiz (Eds.): UCAmI 2023, LNNS 841, pp. 67–77, 2023.
https://doi.org/10.1007/978-3-031-48590-9_7

and diagnosis (FDD) [14]. Recent studies have focused on integrating DL techniques and Conditional Generative Adversarial Networks (CGAN) to improve the prediction and estimation of PV power generation. By effectively harnessing DL methodologies such as recurrent neural networks (RNN) [1,20] and convolutional neural networks (CNN) [2,13,19,22] in combination with Long Short-Term Memory (LSTM), researchers aim to capture the intricate spatio-temporal dynamics inherent in PV systems. Complementing DL, CGANs provide a robust framework to generate realistic synthetic data and effectively differentiate between authentic and fabricated power generation samples [6,16,17]. The convergence of these methodologies holds significant potential to enhance the accuracy and efficiency of estimating PV power output, empowering informed maintenance decisions and optimizing system performance. Exploring DL and CGAN in renewable energy research [7,9] contributes to expanding the existing knowledge base, propelling advancements in PV power forecasting, nowcasting, and system management [10,18,21] into new scientific understanding and practical implementation frontiers.

This work focuses on evaluating the application of Deep Learning (DL) techniques and Conditional Generative Adversarial Networks (CGANs) in nowcasting the output power generation of photovoltaic (PV) systems. Nowcasting refers to the estimation of future power generation based on historical data. Accurate power output prediction is crucial for maintenance and error detection in PV systems.

The main objectives of this work are as follows:

- Evaluate Convolutional Neural Networks (CNNs) combined with Long Short-Term Memory (LSTM) networks as a baseline model for nowcasting power output generation.
- Highlight the relevance of CGAN models in developing estimators and discriminators for PV systems, particularly in maintenance and error detection.
- CGAN model approaches regression problems, employing binary classification of real and fake samples in the discriminator model.

The rest of the paper is organized as follows: Sect. 2 describes the preprocessing of data, the CNN+LSTM baseline model used for nowcasting power output generation, and the CGAN model proposed in this work. Section 4 concludes the paper and discusses the ongoing work.

2 Methods

In this section, we describe the preprocessing of data and the proposed CGAN model to nowcast the output power generation of PV systems. First, data preprocessing for computing features from a sliding window is described. Second, a DL model based on CNN+LSTM is introduced as a baseline to compose advanced CGAN models to nowcast and discriminate false output energy generation in PV systems.

2.1 Segmentation to Nowcast Power Generation

Following a formal definition, a sensor s collects data in real-time in the form of a pair $\overline{s_i} = \{s_i, t_i\}$, where s_i represents a given measurement and t_i the timestamp. Thus, the sensor source data stream s is defined by $\overline{S_s} = \{\overline{s_0}, \ldots, \overline{s_i}\}$ and a given value in a timestamp t_i by $S_s(t_i) = s_i$. In this work, the irradiance on the PV surface G_I, the ambient temperature T_{am}, and the power generation from the PV output P_A provide three data streams that describe the behaviour and energy production of the PV system.

We define several symmetric temporal sliding windows to homogenize the data collected by the different sensors. They are defined by the window size of a time interval $W_w = [W_w^-, W_w^+]$, segment the samples of a given sensor stream $\overline{S_s}$ and aggregate the values $\overline{s_i}$ employing an aggregation function $T_t(S_s, W_w, t^*)$:

$$T_t(S_s, W_w, t^*) = \bigcup_{s_i}^{\overline{s_i}} s_i, t_i \in [t^* - W_w^-, t^* - W_w^+] \tag{1}$$

So, the aggregation $\bigcup_{s_i}$ from the sensor data s_i applied over a short time interval $W_w = [W_w^-, W_w^+]$ represents the relevant value in a given point of time t^*.

Using data aggregation, the signal segmentation is defined by several sliding temporal windows of short size, which are defined by the temporal granularity Δ. The aggregation function aggregates the data in the temporal window. In concrete, the data within each temporal window are averaged $\bigcup = \mu$ for each short-term temporal window within the segment defined in the time interval. The average provides a strong aggregation function to homogenize heterogeneous raw sensor sources in case different collection rates provide them.

So, we obtain a data sequence for each sensor source, whose sequence size is the same for all sources S_s. In this work, two datasets with different input signals have been included for evaluation. In Sect. 3, we describe inputs, window size and time steps set for each one.

2.2 Baseline DL Model for Nowcasting Output Power Generation

This section describes the baseline model for nowcasting output power generation under a DL approach. A given number of sensors configures the model from input signals of sensors S (irradiance on the PV surface and ambient temperature) within a temporal window of W steps configuring an input matrix of SxW. Three layers of 1D CNN are firstly integrated as spatial feature extractors. Next, two layers of LSTM model the temporal dependencies from the upper features of CNN. The combination of CNN-LSTM Hybrid Networks has been selected to provide encouraging results in power consumption [12]. In Fig. 1, we describe the configuration and layers for the proposed model.

The model proposed here has been demonstrated to have a suitable configuration which improves the results from CNN and LSTM in an isolated way [4].

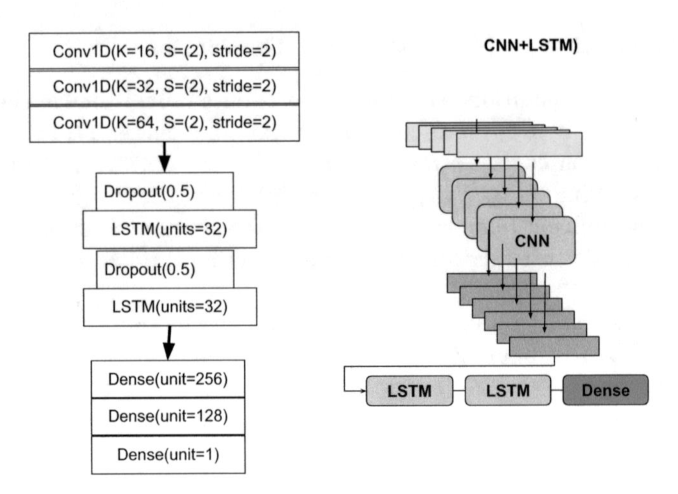

Fig. 1. CNN+LSTM baseline model

2.3 CGAN Regression+Classification for Nowcasting and Discriminating Output Power Generation

From the baseline DL model described in the previous Section, we develop a CGAN model to nowcast output power generation of PV Systems. In the configuration of CGAN models, a generator learns to generate new synthetic data, which are more indistinguishable from real ones; the opposite discriminator learns to distinguish between synthetic and real data. Learning a discriminator from real or fake samples in PV systems includes a straightforward motivation to develop a component for maintenance purposes or fraud detection.

The generator and discriminator are connected in an upper model (CGAN) and are trained adversarially. In learning, the loss function has to follow a zero-sum to encourage the learning of discriminator and generator in a balanced way. The generator develops a regression model for this section, but the discriminator faces to distinguish between real and fake samples on a binary classification problem. We use a similar approach proposed in [3] to develop a suitable model. First, the generator is developed with the baseline model of CNN+LSTM described, whose sensor inputs $S \times W$ are the measurements of S sensor of the PV system in W temporal steps together with a noise input signal N. The output model (energy generation) is connected as input of the discriminator model and the sensor measurements $S \times W$ to distinguish if the energy generation is real or fake. The output of the discriminator is configured as a binary classification with binary cross entropy. In the end, an adversarial model connects both the inputs and output of the generator model to develop a common structure for learning in two steps (which is described in Sect. 3) to balance the following function:

$$E_{x,y}[log(D(x,y))] + E_{x,z}[1 - log(D(x,G(x,z))]$$

Where: i) y represents the ground-truth energy generation from x sensor input, ii) $D(x, y)$ described the discriminator classification of fake and real estimation,

iii) output of $G(x, z)$ represents the estimated energy generation for noise z and samples x. The function derives from the cross-entropy between the real and generated distributions, which is learned to be maximized by the discriminator and minimized by the generator.

In Fig. 2, we provide a visual description of the components and connections which compose this CGAN model.

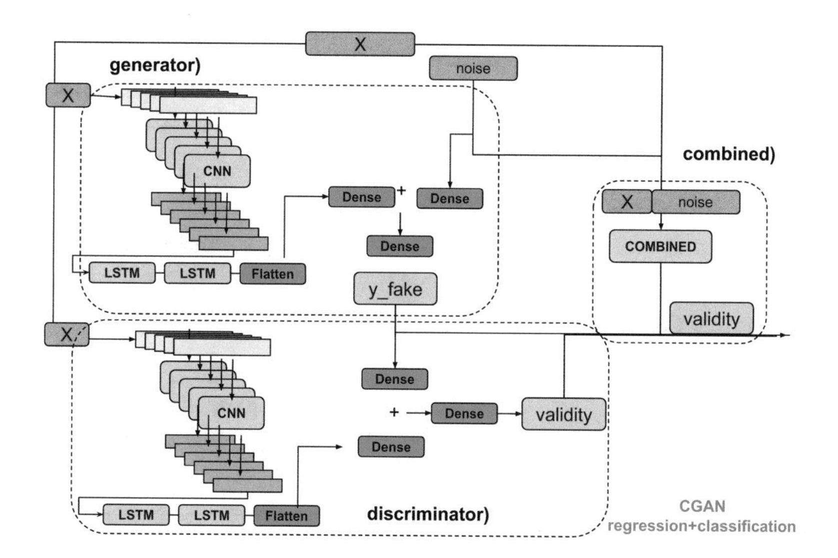

Fig. 2. CGAN regression+classification model

3 Results

This section presents the results from modelling the data collected from two real-world datasets with the proposed adversarial network. The data collected by the photovoltaic system and the code of the models from Opera Project are available in https://github.com/AuroraPR/Opera-UCAmI. Opera is a digital data platform developed by an interdisciplinary team covering the areas of IT, PV, and Electronic Technology, and its objective is to provide O&M (Operations and Maintenance) services for renewable energy installations. This digital platform has been developed with the knowledge and the working data of UniVer, a standard, medium-sized, grid-connected PV system that has been running for the last twenty-five years on the campus of the University of Jaén. Opera platform is also now managing the O&M of this PV system.

3.1 Dataset Description

The segmentation of input and output signals from the PV system is developed for Ambient Temperature and Global Irradiance as the input signals and Output

Power as the output. This process of aggregating and segmenting the data is key in the transfer learning approach since the sample rates generally differ between the domains. In the two contexts of this work, the input data is configured as follows:

- Univer Dataset. The signals are aggregated from the original 30-second sample to a 10-minute average and then segmented in a 90-minute sliding window ($W = 9$). A total of 24 031 samples compose the dataset.
- SolarTech Dataset. The signals are aggregated from the original 1-minute sample to a 10-minute average and then segmented in a 90-minute sliding window ($W = 9$). A total of 17.136 samples compose the dataset.

Using 90 min has developed a suitable window size configuration [4]. In the different training processes for the CGAN models, we have included the same configuration of epochs = 100 and batch = 64 to homogenize results.

3.2 Baseline DL Model

This section describes the results of the baseline DL model to nowcast output power generation. First, we provide a short pseudo-code that illustrates the training process and compares it with CGAN training in the following sections.

Algorithm 1. Baseline training

1: $N \leftarrow len(X)$
2: $batch_size \leftarrow N/batchs$
3: **for** $e \in [0, epochs]$ **do**
4:     **for** $b \in [0, batchs]$ **do**
5:         $i = ranInd(batch_size, N)$ ▷ Random indexes between [0,N]
6:         $x, y \leftarrow X[i], Y[i]$
7:         $model.fit(x, y)$
8:     **end for**
9: **end for**

Next, in Table 1, we present the result of normalized RMSE for each dataset, and in Fig. 3, we present the evolution of evaluation and training loss from datasets SolarTech and UNIVER. To add a frame of reference for these RMSE values, the peak power of the Univer system is 100kWp, so being able to nowcast the output power of the system with an error of 610.85W means that it is possible to provide O&M services (such as detecting system deterioration or malfunction) with a very high level of confidence, by comparing both the expected and the actual system output.

Table 1. Summary of error metrics for Baseline DL model

Dataset	RMSE (W)	NRMSE (W)
SOLAR	6.22	$2.56 \cdot 10^{-2}$
UNIVER	628.72	$2.20 \cdot 10^{-2}$

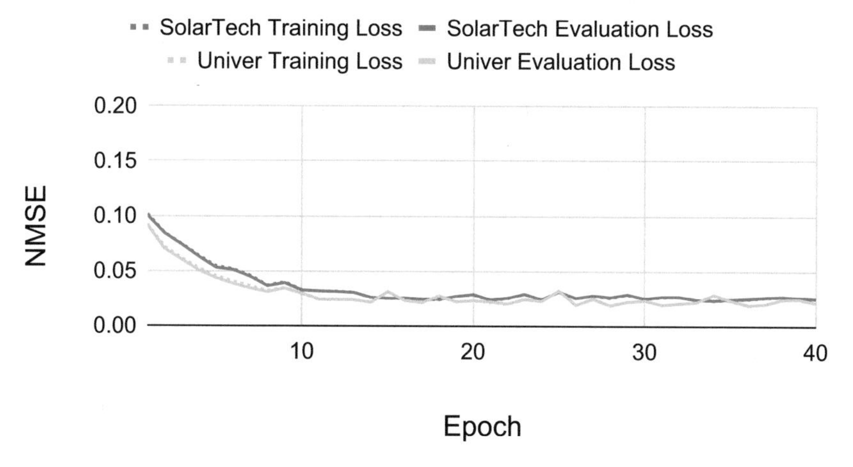

Fig. 3. CNN+LSTM in baseline model) Evaluation and training loss from datasets SOLAR and UNIVER

3.3 CGAN Regression+Classification for Nowcasting and Discriminating

In this section, we present the results provided by the model of CGAN regression+classification, which nowcasts and discriminates potential fake estimations. The model has been trained according to the next configurations:

- Generator noise $noise_{generator}$. It includes noise between $[0, 1]$ based on a random uniform distribution $U(0, 1)$.
- Fusing noise in the generator. A dense layer (64 units) is connected to generator noise to combine features from the baseline model (CNN+LSTM) and generate a pseudo-stochastic output.
- Freezing discriminator layers. In the combined training model, the discriminator layers are frozen to avoid the double training regarding the generator.
- Discriminator noise $noise_{discriminator}$, which generates a fake output generation from prediction. It is key in this work to noise the predicted output of the generator due to the high similarity of the regression of power generation from the baseline model, being non-indiscernible from real power generation. We include a random uniform distribution $U(-0.5, 0.5)$

Next, in Table 2, we present the results of normalized RMSE. Examining the RMSE values in the context of Univer, an error of 513.38W in a 100kWp peak

Algorithm 2. CGAN regression+classification training

```
1: N ← len(X)
2: batch_size ← N/batchs
3: for e ∈ [0, epochs] do
4:     for b ∈ [0, batchs] do
5:         i = ranInd(batch_size, N)                    ▷ Random indexes between [0,N]
6:         x, y_real ← X[i], Y[i]
7:         noise_generator ← U(0, 1)
8:         y_pred = generator.predict(x, noise_generator)
9:         discriminator.train([x, y_real], true)
10:        noise_discriminator ← U(−0.5, 0.5)
11:        y_fake = y_pred + noise_discriminator
12:        discriminator.train([x, y_fake], false)
13:        combined.train([0, x], [y_real, true])
14:    end for
15: end for
```

power PV system implies that the generated data is highly similar to the actual output power of the PV system. In Fig. 4, we present the progression for epochs of evaluation and training loss from datasets SOLAR and UNIVER, as well as the accuracy in true and fake sample prediction for epochs in Solartech and Univer datasets.

Table 2. Summary of error metrics for CGAN regression+classification for nowcasting and discriminating

Dataset	RMSE (W)	NRMSE (W)
SOLAR	5.36	$2.45 \cdot 10^{-2}$
UNIVER	479.66	$1.75 \cdot 10^{-2}$

3.4 Discussion

This section describes the results of evaluating adversarial networks concerning a Deep Learning base model. First, we highlight the considerable improvement in the generator, whose accuracy (measured in RMSE) concerning the baseline system is very relevant in both datasets.

Secondly, we highlight the importance of including an error concerning the generator to simulate false predictions. On this point, a uniform distribution [−0.5,0.5] has been successful. An accuracy of 0.86 in classifying real predictions and 0.8 in fake predictions is remarkable. Here, we note that the balance of these values is the key to improving the generator, having experienced a deterioration when the discriminator increased its performance (including a wider additive error distribution). In this sense, we point out that many of the false predictions

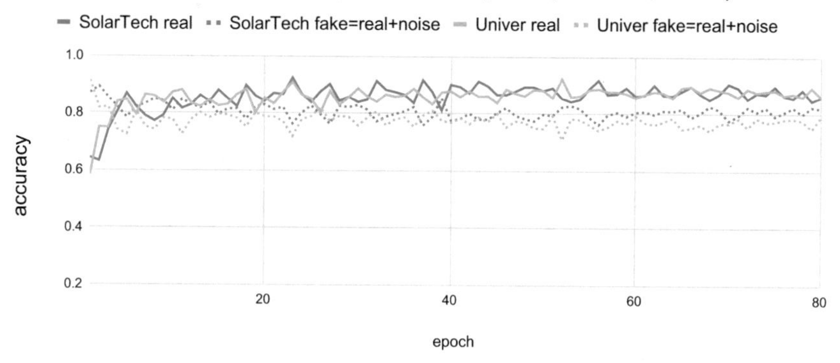

Fig. 4. CGAN regression+classification. Top) progression of Bottom) Accuracy in real and fake sample prediction for epochs and dataset)

would be real (due to a random additive error close to zero), so this classification percentage is in conditions of the high similarity of generated energy.

In this context, this system holds significant promise for fraud detection because the generator is capable of learning to generate samples that closely resemble real data. If the discriminator struggles to distinguish between fake and real samples, it effectively learns from a broad and diverse data set. This characteristic enhances the system's robustness in detecting anomalies and predicting consumption, including subtle variations that may suggest users manipulate energy consumption to their advantage. Essentially, the system's ability to adapt and learn from a wide spectrum of data sources enhances its effectiveness in detecting fraudulent activities and predicting deviations from expected behaviour, thereby making it a valuable tool in fraud prevention and detection within the energy sector.

4 Conclusions and Ongoing Works

In conclusion, this work explores the potential of Deep Learning and Adversarial Networks for nowcasting and discriminating the output power generation in photovoltaic (PV) systems. The researchers begin with a baseline model consisting of Convolutional Neural Networks and Long Short-Term Memories to capture spatiotemporal patterns. Building upon this, they introduce a discriminator and generator using Conditional Generative Adversarial Networks. The adversarial network enables accurate estimation and discrimination of erroneous output power generation.

We have evaluated the proposed approach using two real-world datasets and achieved promising results that directly align with the practical applications of maintenance in PV systems.

In future work, we will focus on modelling a CGAN regression doping model for nowcasting to rectify output power generation from fake estimations.

Acknowledgements. This contribution has been supported by the *Cátedra ELAND for Renewable Energies* of the University of Jaén, by the Spanish government through the project RTI2018-098979-A-I00.

References

1. Abdel-Nasser, M., Mahmoud, K.: Accurate photovoltaic power forecasting models using deep LSTM-RNN. Neural Comput. Appl. **31**, 2727–2740 (2019)
2. Agga, A., Abbou, A., Labbadi, M., El Houm, Y., Ali, I.H.O.: CNN-LSTM: an efficient hybrid deep learning architecture for predicting short-term photovoltaic power production. Electr. Power Syst. Res. **208**, 107908 (2022)
3. Aggarwal, K., Kirchmeyer, M., Yadav, P., Keerthi, S.S., Gallinari, P.: Regression with conditional GAN. arXiv preprint arXiv:1905.12868 (2019)
4. Almonacid-Olleros, G., Almonacid, G., Fernandez-Carrasco, J.I., Quero, J.M.: Opera. dl: deep learning modelling for photovoltaic system monitoring. In: Multidisciplinary Digital Publishing Institute Proceedings. vol. 31, p. 50 (2019)
5. Azuatalam, D., Paridari, K., Ma, Y., Förstl, M., Chapman, A.C., Verbič, G.: Energy management of small-scale PV-battery systems: a systematic review considering practical implementation, computational requirements, quality of input data and battery degradation. Renew. Sustain. Energy Rev. **112**, 555–570 (2019)
6. Bu, X., Wu, Q., Zhou, B., Li, C.: Hybrid short-term load forecasting using CGAN with CNN and semi-supervised regression. Appl. Energy **338**, 120920 (2023)
7. Dong, W., Chen, X., Yang, Q.: Data-driven scenario generation of renewable energy production based on controllable generative adversarial networks with interpretability. Appl. Energy **308**, 118387 (2022)
8. Europe, S.P.: Global market outlook for solar power/2019-2023. Tech. Rep, Solar Power Europe, Brussels, Belgium (2019)
9. He, G., Liu, K., Wang, S., Lei, Y., Li, J.: CWM-CGAN method for renewable energy scenario generation based on weather label multi-factor definition. Processes **10**(3), 470 (2022)
10. Huang, X., et al.: Time series forecasting for hourly photovoltaic power using conditional generative adversarial network and Bi-LSTM. Energy **246**, 123403 (2022)
11. Is, I.R.P.G.: Being turbocharged as countries seek to strengthen energy security
12. Kim, T.-Y., Cho, S.-B.: Predicting the household power consumption using CNN-LSTM hybrid networks. In: Yin, H., Camacho, D., Novais, P., Tallón-Ballesteros, A.J. (eds.) IDEAL 2018. LNCS, vol. 11314, pp. 481–490. Springer, Cham (2018). https://doi.org/10.1007/978-3-030-03493-1_50
13. Li, G., Xie, S., Wang, B., Xin, J., Li, Y., Du, S.: Photovoltaic power forecasting with a hybrid deep learning approach. IEEE Access **8**, 175871–175880 (2020)
14. Mansouri, M., Trabelsi, M., Nounou, H., Nounou, M.: Deep learning-based fault diagnosis of photovoltaic systems: a comprehensive review and enhancement prospects. IEEE Access **9**, 126286–126306 (2021)
15. Massaoudi, M., Chihi, I., Abu-Rub, H., Refaat, S.S., Oueslati, F.S.: Convergence of photovoltaic power forecasting and deep learning: State-of-art review. IEEE Access **9**, 136593–136615 (2021)
16. Mohammadpourfard, M., Ghanaatpishe, F., Mohammadi, M., Lakshminarayana, S., Pechenizkiy, M.: Generation of false data injection attacks using conditional generative adversarial networks. In: 2020 IEEE PES Innovative Smart Grid Technologies Europe (ISGT-Europe), pp. 41–45. IEEE (2020)

17. Pan, Y., Liu, K., Shen, Z., Cai, X., Jia, Z.: Sequence-to-subsequence learning with conditional GAN for power disaggregation. In: ICASSP 2020-2020 IEEE International Conference on Acoustics, Speech and Signal Processing (ICASSP), pp. 3202–3206. IEEE (2020)
18. Peng, Y., Ye, L., Zhao, Y., Li, Z., Wang, X., Li, P.: Stochastic scenario generation for wind power and photovoltaic system based on CGAN. In: 2022 IEEE/IAS Industrial and Commercial Power System Asia (I&CPS Asia), pp. 1441–1446. IEEE (2022)
19. Qu, J., Qian, Z., Pei, Y.: Day-ahead hourly photovoltaic power forecasting using attention-based CNN-LSTM neural network embedded with multiple relevant and target variables prediction pattern. Energy **232**, 120996 (2021)
20. Rajagukguk, R.A., Ramadhan, R.A., Lee, H.J.: A review on deep learning models for forecasting time series data of solar irradiance and photovoltaic power. Energies **13**(24), 6623 (2020)
21. Song, Z., Huang, Y., Li, X., Li, C.: Constructing method of multi-regional photovoltaic power output scenarios based on conditional generative adversarial network. In: 2022 IEEE 6th Conference on Energy Internet and Energy System Integration (EI2), pp. 592–598. IEEE (2022)
22. Zhang, J., Verschae, R., Nobuhara, S., Lalonde, J.F.: Deep photovoltaic nowcasting. Sol. Energy **176**, 267–276 (2018)

Shedding Light on the Energy Usage of Activity Recognition Systems in Homes

Alicia Montoro Lendínez[1](✉), José Luis López Ruiz[1], David Díaz Jiménez[1], Macarena Espinilla Estévez[1], and Chris Nugent[2]

[1] Department of Computer Science, University of Jaén, Jaén, Spain
alicia@gmail.com

[2] School of Computing, Ulster University, Belfast, UK

Abstract. Activity recognition systems are composed of devices with sensors which, through artificial intelligence techniques, can detect activities performed by people in their homes. Many of these systems are deployed in multi-occupancy environments, so indoor localization approaches are combined with activity recognition systems to achieve discrimination of activities in the same space. The benefits of these systems are numerous such as remote monitoring for anomaly warning or improved safety of the monitored person. Although there is an extensive study on these systems from the technical point of view, there is an important gap in the literature on their energy consumption. This fact is even more relevant considering that one of the most important concerns in society is the prices of electricity and it has had a great variability, with increases, due to the pandemic and the war in Ukraine. This work aims to address this scientific gap through the energy evaluation of a home activity recognition system in two scenarios. First, an ambient intelligence apartment (Smart Lab) of the University of Jaén and, then, a single-family house. The evaluation carried out provides quantitative data considering the data of the year 2022, with a price between 3.125€and 4.018€and qualitative data from the point of view of patients, healthcare professionals and researchers.

Keywords: Activity recognition systems · Indoor localization · Energy consumption · Remote monitoring · Sustainable Development Goals

1 Introduction

Activity recognition systems make it possible to know the different activities carried out by one or several monitored people in a specific environment [4]. The advantages provided by activity recognition systems are multiple, such as the detection of certain irregularities in the monitored people's activity (falls, lack of control in the schedule of meals or daily hygiene) [10], help to comply with

Grant PID2021-127275OB-I00 funded by MCIN/AEI/10.13039/501100011033 and by "ERDF A way of making Europe".

J. Bravo and G. Urzáiz (Eds.): UCAmI 2023, LNNS 841, pp. 78–89, 2023.
https://doi.org/10.1007/978-3-031-48590-9_8

healthy habits (adequate rest time or physical activity) [9], improve the quality of life of people with a certain degree of physical or cognitive dependence [7] and guarantee their safety and the family members' confidence [22].

Within activity recognition, there is special interest in approaches aimed at detecting activities in multi-occupancy environments [13, 19], i.e. in environments where several inhabitants live and it is necessary to discriminate which activity is carried out by which inhabitant. In multi-occupancy contexts, indoor location systems have played a crucial role in recent years [14, 24]. In this way, when an event is generated by a sensor in the home, the person closest to that sensor is located, associating that event with the interaction produced by the inhabitant in the home. Inevitably, combining indoor location systems for multi-occupancy activity recognition systems requires a larger number of devices and thus higher energy consumption. Among the multi-occupancy activity recognition systems, the ACTIVA system [15, 18] is of particular interest, as it has been tested as a suitable system for activity recognition in elderly people's nursing homes in order to monitor their inhabitants. In addition, the ACTIVA system has the ACTIVA app that allows to visualize the activity of each user in real time and to display notifications for the caregivers of the elderly.

In the literature we can find a wide range of multi-occupancy human activity recognition systems, which evaluate various measures of accuracy, multiple algorithms or various types of sensors, among others [6, 20]. However, until now, no study analysing the energy consumption of activity recognition systems can be found in the scientific literature. The study of the energy consumption of this type of system is very important for its democratisation and viability in the future, since energy prices increase is one of the main problems of the citizens in developed countries [12]. Moreover, this fact becomes even more relevant when the population with the greatest energy concerns includes the elderly, the target population of the vast majority of activity recognition systems [3]. In the context of energy consumption, another important factor to take into account is the current variability of energy prices due to the past pandemic, the current crisis in Ukraine and other factors such as the rising cost of gas, the demand and the cost of carbon dioxide emissions[11] and the fulfilment of the 2030 agenda through the Sustainable Development Goals (SDGs). Specifically, the ACTIVA system is fully aligned with SDG7 in energy and SDG3 in health systems [16, 23].

This work aims to shed light on the energy consumption of a multi-occupancy activity recognition system based on indoor location systems. For this purpose, the ACTIVA [15, 18] system is deployed in two multi-occupancy environments. On the one hand, in the ambient intelligence flat of the University of Jaén called UJAmI [8] and, on the other hand, in a single-family house with 4 inhabitants. Smart plugs to measure consumption are placed in the ACTIVA system devices in order to measure their consumption in the two scenarios. Finally, based on the energy cost in Spain in 2022, the maximum and average cost of this system in the two proposed scenarios are determined in order to provide quantitative data on its use.

The structure of the paper is as follows. Section 2 reviews the indoor location-based multi-occupancy activity recognition system called ACTIVA and presents the existing tools on the market to measure energy consumption. In Sect. 3 the two scenarios where such a system will be deployed are proposed. Then, Sect. 4 presents the obtained cost of the energy consumption, the material cost and the total cost in the two evaluated scenarios according to the price of electricity in the year 2022. Finally, the conclusions are presented in Sect. 5.

2 Materials and Methods

This section presents the ACTIVA system and the different devices available on the market to measure energy consumption.

2.1 ACTIVA System

The ACTIVA [15,18] system is characterised by discriminating multi-occupancy activities with a BLE indoor location approach. The activities that will be monitored in the ACTIVA system in the two scenarios will be the following: daily physical activity, taking medication, sleeping and grooming including showering and brushing teeth. The devices which are integrated in the ACTIVA system and which generate the events according to the interactions of the inhabitants in the home are the following:

- Open/close sensor[1]. Installed both on the main door of the house to know the entrances and exits of the monitored inhabitant and on a box for medicines in order to know the intake of medicines by the monitored inhabitant. It is powered by a CR1632 battery, lasts approximately 1 year and its cost per battery unit is 0.57cts.
- Motion sensor[2]. Installed both in the headboard of the bed to know the time of rest of the monitored inhabitant and in the bathroom focused on toothbrushes to know the daily cleanliness of the monitored inhabitant. It is powered by two CR2450 batteries, lasts approximately 5 years and i its cost per battery unit is 0.90cts.
- Temperature, humidity and pressure sensor[3]. Installed in the bathroom near the shower area to know when the monitored inhabitant showers. It is powered by a CR2032 battery, lasts approximately 2 years and its cost per battery unit is 0.93cts.

In multi-occupancy environments it is necessary to associate each event with the inhabitant who has carried out the interaction that generates the event. For this reason, the ACTIVA system incorporates a location system through RSSI values of the BLE protocol. For this purpose, each inhabitant wears an activity wristband[4] and several Raspberry Pi 4B devices, as BLE beacons, are playing

[1] https://www.aqara.com/eu/door_and_window_sensor.html.
[2] https://www.aqara.com/eu/product/motion-sensor-p1.
[3] https://www.aqara.com/eu/temperature_humidity_sensor.html.
[4] https://www.mediamarkt.es/es/product/_pulsera-de-actividad-xiaomi-mi-band-3-oled-puls%C3%B3metro-sensor-frecuencia-card%C3%ADaca-negro-1434035.html.

the role of anchor. These Raspberry Pi anchor are distributed in the rooms where you want to locate the inhabitant and thanks to the BLE takes the RSSI values of the wristband every second. Additionally, one of the BLE anchors plays the role of a central node and is responsible for receiving the events generated by the sensors and the RSSI flows from each of the anchors. This central node is responsible for sending all the information to a server in the cloud which infers in real time, on the one hand, the location of each inhabitant and, on the other hand, the activities under study according to the events generated. In addition, the storage, persistence and visualisation of the data received and the knowledge inferred is carried out on the central server.

Since the purpose of this paper is focused on the energy consumption of the devices of the indoor location-based multi-occupancy activity recognition system, the artificial intelligence-based methods for knowledge inference, which are based on the location proposal of the authors López-Medina et al.[14] and Albín-Rodríguez et al. [2], are not described in detail.

2.2 Measuring Energy Consumption

This section reviews alternatives for measuring home energy consumption, which are listed below.

- Energy consumption meters. These are devices which are simply plugged into a plug or household appliance and measure their consumption in real time. Normally, they have a screen from which the different variables measured can be displayed. Some examples that can be found on the market Belkin Converse Insight[5] or Kill A Watt[6].
- Smart energy meters. These are more advanced devices than the previous ones. They are connected to the electricity grid and measure energy consumption in real time and, in addition, all the information can be viewed from a website or app. In this way the user can take measures to reduce consumption. Several examples are compared in [1] that try to control the whole home and with artificial intelligence give measures on the user's behaviour to reduce energy consumption in the home. Another example is the smart plugs that have been used in this work Tapo P100[7].
- Energy audits. These are assessments of energy consumption in certain environments. The assessment is carried out by a professional who helps to identify areas where energy consumption can be reduced and what actions can be taken to do so [21].

As mentioned above, the tool for measuring energy consumption in this case study was smart plugs. Specifically, Tapo P100[8] smart plugs have been chosen as they allowed us to visualise the data in real time from the commercial application and to export the data in .csv format for analysis. Figure 1a shows the start

[5] https://www.belkin.com/uk/support-article/?articleNum=5381.
[6] http://www.p3international.com/products/p4460.html.
[7] https://www.tapo.com/es/product/smart-plug/tapo-p100/.
[8] https://www.tapo.com/es/product/smart-plug/tapo-p100/.

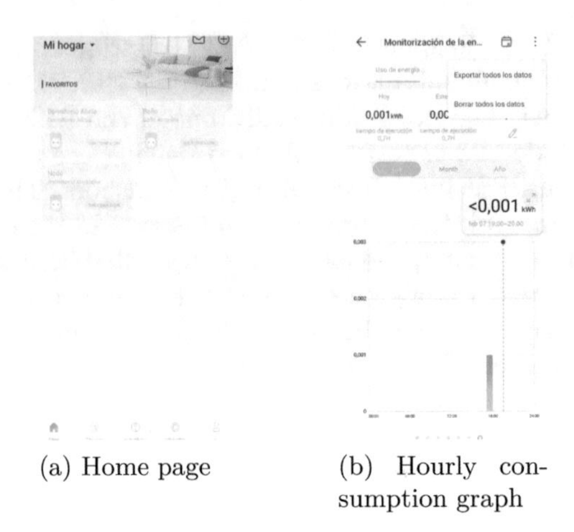

(a) Home page (b) Hourly consumption graph

Fig. 1. Tapo app visualisation.

screen of the application with the current devices in operation and Fig. 1b shows the monitoring of the hourly energy consumption of one of the smart plugs. It should be taken into account that a smart plug was installed at each node and anchor, on the other hand, the sensors have a very low power consumption and are powered externally by button type batteries, so measuring their consumption is not the cost of the inhabitants.

3 Evaluated Scenarios

This section presents the two scenarios and their characteristics. In both scenarios, the multiple devices of the ACTIVA system will be installed.

3.1 Smart Lab

The Smart Lab has a total surface area of $25.44m^2$ which is divided into a small hall, a kitchen, a living room with an office and a bedroom with a bathroom inside.

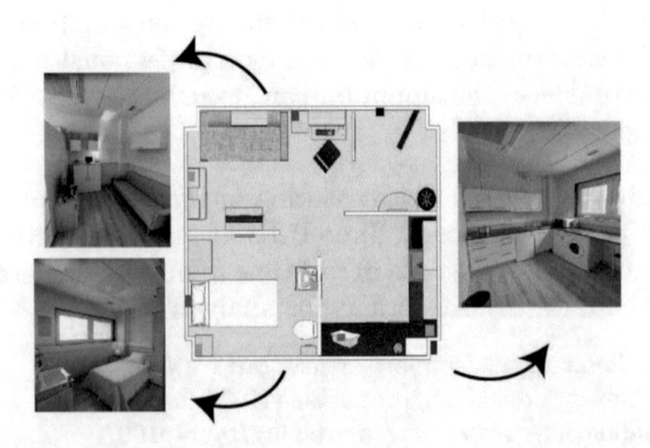

Fig. 2. Layout of the Smart Lab with the distribution of the devices.

In Fig. 2 shows the layout of the Smart Lab of the UJA [8] together with the different devices that have been installed, which are listed below: central node (red), living room anchor (blue), kitchen anchor (blue), bedroom/bathroom anchor (blue colour), motion sensor for brushing (green), motion sensor for resting (purple), temperature, humidity and pressure sensor for the bathroom (yellow), open/close sensor for the medication box (pink) and open/close sensor for the main door (black).

In particular, the central node and the kitchen and living room anchors were fitted with a cooling system. This cooling system includes fan and heatsink. Only, the fan was not installed in the bedroom/bathroom anchor to avoid disturbance and interruption of the inhabitant's rest. The anchors or nodes that include this complete cooling system will have an increase in energy consumption depending on the time the fan is active. Although all fans have been programmed to only turn on if the temperature of the CPU of the Raspberry Pi 4B exceeds 70 °C and turn off when the temperature of the CPU is below 60 °C [5,17], preventing the fan has an unnecessary and excessive consumption.

3.2 Single-Family House

The single-family house consists of two floors. The ground floor has a surface area of 120 m^2 and consists of a small hall, a living room, a toilet, a kitchen with utility room and an outside terrace (refer to Fig. 3).

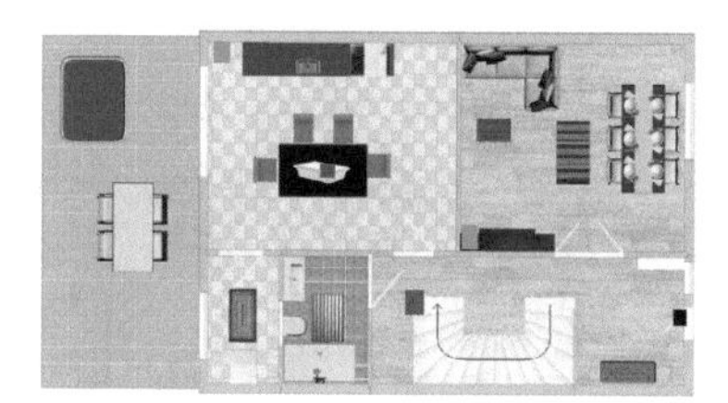

Fig. 3. Ground floor plan of the single-family house with the distribution of the devices.

The first floor has a surface area of 97 m^2 and comprises three single bedrooms, a bathroom and a double bedroom with its own small bathroom (see Fig. 4). Although four people live in the single-family house, only two will be monitored as in the Smart Lab. Also, in Fig. 3 and Fig. 4 it can be shown the different devices that have been installed in both floors.

The ground floor hosts the central node (red), the living room anchor (blue), the kitchen anchor (blue), open/close sensor for the medication box (pink) and open/close sensor for the main door (black). On the first floor there is the bedroom anchor (blue), the bathroom anchor (blue), the motion sensor for brushing (green), the motion sensor for resting (purple) and the temperature, humidity and pressure sensor for the bathroom (yellow). As with the Smart Lab, the fan has not been installed in the bedroom anchor.

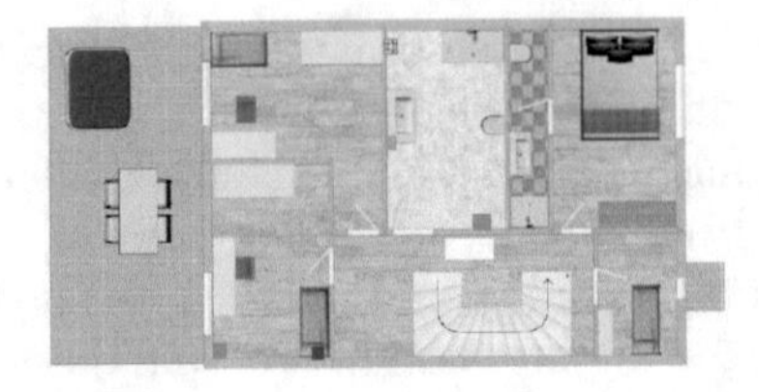

Fig. 4. First floor plan of the single-family house with the distribution of the devices.

4 Results

This section presents the cost results for energy consumption cost, material cost and total cost.

4.1 Energy Consumption Cost

To obtain energy cost data, the data on the average price per €/kWh was obtained from the TarifaLuzHora[9] website. This website publishes data on the Spanish electricity grid with prices referring to the regulated market and on this data the most expensive day, the most expensive week and the most expensive month could be obtained. Figure 5 shows the history over 2022 of the average monthly price of €/kWh obtained from this website.

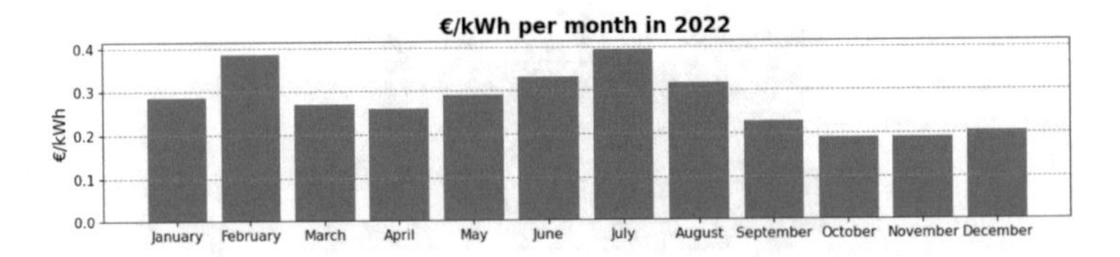

Fig. 5. Average historical data €/kWh in 2022.

Table 1 shows the quantitative consumption data for each device of the ACTIVA System in the Smart Lab scenario and Table 2 for the single-family house scenario.

The first column of Table 1 and Table 2 shows the devices displayed, the second column shows whether the fan is installed and the third column shows the average daily energy consumption in kWh for each device. To obtain it, the total energy consumption obtained during the experiment according to each anchor or sink is divided by the 21 days of the experiment and the following results are obtained. Then, in the following columns, the energy consumption for the most expensive day of 2022, the most expensive week of 2022 and the most expensive month of 2022 are presented in three sections. For these three

[9] https://tarifaluzhora.es.

Table 1. Dialy, weekly, monthly energy consumption from Smart Lab.

Device	With fan	Daily energy consumption (kWh)	Most expensive day in 2022 "8/03/2022"		Most expensive week in 2022 "7/03/2022–13/03/2022"		Most expensive month in 2022 "August"	
			Price for this day (€/kWh)	Total (€)	Price for this week (€/kWh)	Total (€)	Price for this month (€/kWh)	Total (€)
Kitchen Anchor	Yes	1.2915/21 = 0.0615	0.71533	0.044	3.4823	0.214	12.25216	0.754
Living Room Anchor	Yes	1.2915/21 = 0.0615	0.71533	0.044	3.4823	0.214	12.25216	0.754
Bedroom/Bathroom Anchor	No	1.2705/21 = 0.0605	0.71533	0.043	3.4823	0.211	12.25216	0.741
Sink	Yes	1.5015/21 = 0.0715	0.71533	0.051	3.4823	0.249	12.25216	0.876
			Total for the most expensive day in 2022 (€): 0.182		**Total for the most expensive week in 2022 (€): 0.931**		**Total for the most expensive month in 2022 (€): 3.125**	

cases, the price per kWh in €/kWh and the total price per device in €/kWh are shown in two separate columns. And finally, at the bottom, the total price of the entire system with all devices is presented in €.

Table 2. Dialy, weekly, monthly energy consumption from single-family house.

Device	With fan	Daily energy consumption (kWh)	Most expensive day in 2022 "8/03/2022"		Most expensive week in 2022 "7/03/2022–13/03/2022"		Most expensive month in 2022 "August"	
			Daily price (€/kWh)	Total (€)	Weekly price (€/kWh)	Total (€)	Monthly price (€/kWh)	Total (€)
Kitchen Anchor	Yes	1.3608/21 = 0.0648	0.71533	0.046	3.4823	0.226	12.25216	0.793
Living Room Anchor	Yes	1.3608/21 = 0.0648	0.71533	0.046	3.4823	0.226	12.25216	0.793
Bedroom Anchor	No	1.3398/21 = 0.0638	0.71533	0,046	3.4823	0.222	12.25216	0.782
Bathroom Anchor	Yes	1.4406/21 = 0.0686	0.71533	0.049	3.4823	0.24	12.25216	0.84
Sink	Yes	1.3818/21 = 0.0658	0.71533	0.047	3.4823	0.23	12.25216	0.81
			Total for the most expensive day in 2022 (€): 0.234		**Total for the most expensive week in 2022 (€): 1.144**		**Total for the most expensive month in 2022 (€): 4.018**	

It is clear that the daily consumption (expressed in kWh) of the devices installed in the Smart Lab and in the single-family house are different. The main determining factors for these results to be different are as follows:

- The cooling system. Not all devices have the active cooling system (fan) installed. For example, in the case of the bedroom anchor, the fan is not installed so as not to interrupt the user's rest although the use of the fan avoids any overheating of the devices and possible breakage. The central node and the anchors that do have a fan will have a higher energy consumption depending on the operating time of the fan.
- Ambient temperature and ventilation. The Smart Lab is a study scenario where the temperature is kept lower and more stable (above 19 °C) while in the single-family house, depending on the room, there will be different temperatures. For example, the room with the highest temperature (above 25 °C) and the lowest ventilation is the bathroom. Therefore, the fan of this device installed in the bathroom has to work harder and therefore the energy consumption is higher.
- The data flow. While the anchors are simply responsible for obtaining the RSSI value to obtain the location of the inhabitant, the central node is responsible for receiving the data obtained by the anchors and the different sensors

installed and sending all the information to the server in the cloud. Consequently, the central node carries out a greater processing of the information and this greater work on the CPU affects energy consumption, which is higher in the central node.

Having examined the different constraints affecting the energy consumption of the various devices, some of the results obtained are specifically discussed.

In the Smart Lab results, a scenario with ideal conditions, it can be seen that the consumption is higher in the central node (0.0715 kWh) than in the anchors. This is due to the fact that the data processing and CPU work in the central node is higher and, also, the fan has to work harder. Among the anchors, the one that consumed the least was the bedroom/bathroom (0.0605 kWh) due to not having the fan installed. Thus, the system made up of one node and three Smart Lab anchors has a consumption price of 3.125 € per month, considering the most expensive month of 2022.

On the other hand, to analyse the results of the single-family house, which is the scenario without ideal conditions, the different temperatures of the rooms must be taken into account. This will cause the fan to work more or less and therefore the devices will consume more or less energy. The kitchen, bedroom, corridor and living room have a temperature of 19 $°C$ and the bathroom has a temperature of 25 $°C$. In the rooms at 19 $°C$, the consumption of the anchors is 0.0648 kWh and the consumption of the central node is 0.0658 kWh. However, on this occasion the central node does not have the highest consumption because in the bathroom room there is an anchor that consumes more, 0.0686 kWh. Again, it is the bedroom anchor, without the fan installed, that consumes the least of all the devices deployed, (0.0638 kWh). So the system consisting of one node and four anchors in the single-family house has a consumption price of 4.018 € per month, considering the most expensive month of 2022.

Energy Consumption Cost Survey. In order to ensure economic viability in relation to energy consumption, a survey was carried out among different profiles with different points of view, such as 10 researchers (4 men and 4 women between 25–30 years, 1 woman between 30–40 years and 1 man between 40–50 years), 6 health professionals (2 men between 30–40 years, 2 men and 1 woman between 40–50 years and 1 man between 50–60 years) and 10 users as elderly people (2 men and 1 woman between 60–65 years, 1 man and 1 woman between 65–70 years and 3 men and 2 women between 70–75 years). The survey's questions was: "Do you consider it appropriate to pay a monthly fee of 4 €as part of the electricity costs associated with the installation of the ACTIVA system in your home? A Likert scale survey was used to collect their opinions on this question, and examples of energy consumption were provided, e.g. the monthly consumption of a 22 kWh television is equivalent to 275.48 €in August 2022". The results of this survey are presented in Fig. 6, where it can be seen that the three profiles are in greater percentage in agreement with this fee. Specifically, researchers 69%, health professionals 75% and users 60%.

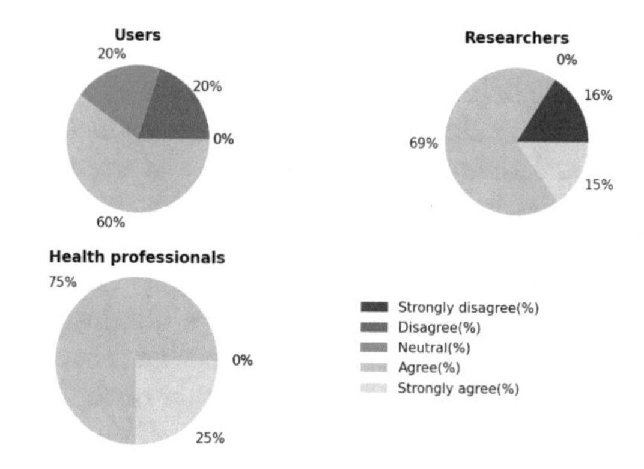

Fig. 6. Results from the energy consumption cost survey

4.2 Material Cost

The material cost of the system will depend on the deployment of devices and sensors. Table 3 presents the material cost in the Smart Lab and in the single-family house.

Table 3. Material cost in Smart Lab and single-family house scenario.

Material	Price per unit (€)	Units in Smart Lab scenario	Units in single-family house scenario
Raspberry Pi Kit	130.65	4	5
Motion sensor + CR2032 batteries	28.99	2	2
Open/Close sensor + CR1632 batteries	19.99	2	2
Temperature, humidity and pressure sensor + CR2450 batteries	22.99	1	1
Conbee II	30.46	1	1
Bluetooth USB	12.99	4	5
Wristband Xiaomi Mi Band 3	55.43	1	1
		Total: 781.4€	**Total: 925.04€**

5 Conclusions and Future Lines

Up to the present, there are no proposals in the literature that provide quantitative data on the energy cost of activity recognition systems based on indoor location system. This paper has described two scenarios to calculate the energy cost of the ACTIVA system: the Smart Lab of the University of Jaén (UJAmI) and a single-family house. For both scenarios, quantitative data has been obtained regarding energy consumption taking into account historical data on the cost of energy during the year 2022 (most expensive day, most expensive week and most expensive month of 2022). Subsequently, the analysis concluded that the active cooling system (fan) and the ambient temperature of the rooms chosen for the installation of the devices have a strong influence on consumption. Therefore, in the Smart Lab, being a scenario with more favourable conditions, the energy consumption data obtained are lower than in the single-family house. However, for both scenarios, the energy consumption prices obtained are affordable and economic, ranging between 3.125 € and 4.018 € per month, considering the most expensive month of 2022. In addition, a calculation of the material cost of the ACTIVA system in the two scenarios was also carried out.

Finally, it is important to note that the study developed in this work in two scenarios (Smart Lab and single-family house) is only an initial pilot. However, this initial pilot aims to shed light on a new line of research such as energy consumption studies in smart systems based on location and recognition of daily life activities. This type of energy consumption studies could be carried out in different scenarios and with different people monitoring projects in order to align these smart systems with the sustainable development goals of the 2030 agenda.

References

1. Alahmad, M.A., Wheeler, P.G., Schwer, A., Eiden, J., Brumbaugh, A.: A comparative study of three feedback devices for residential real-time energy monitoring. IEEE Trans. Ind. Electron. **59**(4), 2002–2013 (2012). https://doi.org/10.1109/TIE.2011.2165456
2. Albín-Rodríguez, A.P., De-La-Fuente-Robles, Y.M., López-Ruiz, J.L., Verdejo-Espinosa, Á., Espinilla Estévez, M.: Ujami location: a fuzzy indoor location system for the elderly. Int. J. Environ. Res. Public Health **18**(16), 8326 (2021)
3. Arenas Pinilla, E.M., et al.: La pobreza energética en españa (2019)
4. Arshad, M.H., Bilal, M., Gani, A.: Human activity recognition: review, taxonomy and open challenges. Sensors **22**(17), 6463 (2022)
5. Benoit-Cattin, T., Velasco-Montero, D., Fernández-Berni, J.: Impact of thermal throttling on long-term visual inference in a CPU-based edge device. Electronics **9**(12), 2106 (2020)
6. Bibbò, L., Carotenuto, R., Della Corte, F.: An overview of indoor localization system for human activity recognition (HAR) in healthcare. Sensors **22**(21), 8119 (2022)
7. Chen, D., Bharucha, A.J., Wactlar, H.D.: Intelligent video monitoring to improve safety of older persons. In: 2007 29th Annual International Conference of the IEEE Engineering in Medicine and Biology Society, pp. 3814–3817. IEEE (2007)
8. Espinilla, M., Martínez, L., Medina, J., Nugent, C.: The experience of developing the UJAMI smart lab. IEEE Access **6**, 34631–34642 (2018)
9. Fleury, A., Vacher, M., Noury, N.: SVM-based multimodal classification of activities of daily living in health smart homes: sensors, algorithms, and first experimental results. IEEE Trans. Inf Technol. Biomed. **14**(2), 274–283 (2009)
10. Gannapathy, V.R., Ibrahim, A., Zakaria, Z.B., Othman, A.R.B., Latiff, A.A.: Zigbee-based smart fall detection and notification system with wearable sensor (e-safe). Int. J. Res. Eng. Technol. **2**(8), 337–344 (2013)

11. Jääskeläinen, J., Huhta, K., Syri, S.: The anatomy of unaffordable electricity in northern Europe in 2021. Energies **15**(20), 7504 (2022)
12. Lepetit, N.G., Biard, E., Aparisi-Cerdá, I., Brazzini, T., Montagud, C., Gómez-Navarro, T.: Measuring the discomfort of energy vulnerable elderly people: recommendations for solutions. In: IOP Conference Series: Earth and Environmental Science, vol. 1085, p. 012016. IOP Publishing (2022)
13. Li, Q., Gravina, R., Li, Y., Alsamhi, S.H., Sun, F., Fortino, G.: Multi-user activity recognition: challenges and opportunities. Inf. Fusion **63**, 121–135 (2020)
14. López-Medina, M., Espinilla, M., Cleland, I., Nugent, C., Medina, J.: Fuzzy cloud-fog computing approach application for human activity recognition in smart homes. J. Intell. Fuzzy Syst. **38**(1), 709–721 (2020)
15. López Ruiz, J.L., Espinilla Estévez, M., Medina Quero, J., Verdejo Espinosa, M.Á., Salguero Hidalgo, A.G.: Aplicación móvil activa (2021). https://www.safecreative.org/work/2111159810407
16. López, J.L., Espinilla, M., Verdejo, Á.: Evaluation of the impact of the sustainable development goals on an activity recognition platform for healthcare systems. Sensors **23**, 3563 (2023)
17. Machowski, J., Dzieńkowski, M.: Selection of the type of cooling for an overclocked raspberry pi 4b minicomputer processor operating at maximum load conditions. J. Comput. Sci. Inst. **18**, 55–60 (2021)
18. Martínez, J.M.M., et al.: Sistema inteligente de reconocimiento de actividades en el entorno de envejecimiento activo y de seguridad (activa): evaluación de herramienta informática desde el ámbito social. In: Conocimientos, investigación y prácticas en el campo de la salud: actualización de competencias, pp. 115–120. Asociación Universitaria de Educación y Psicología (ASUNIVEP) (2021)
19. Mohamed, R., Perumal, T., Sulaiman, M.N., Mustapha, N.: Multi resident complex activity recognition in smart home: a literature review. Int. J. Smart Home **11**(6), 21–32 (2017)
20. Ramasamy Ramamurthy, S., Roy, N.: Recent trends in machine learning for human activity recognition—a survey. Wiley Interdisc. Rev. Data Mining Knowl. Disc. **8**(4), e1254 (2018)
21. Rey Martínez, F.J., Velasco Gómez, E.: Eficiencia energética en edificios. Certificación y auditorías energéticas: certificación y auditorías energéticas. Ediciones Paraninfo, SA (2006)
22. Schrader, L., et al.: Advanced sensing and human activity recognition in early intervention and rehabilitation of elderly people. J. Popul. Ageing **13**, 139–165 (2020)
23. Ángeles Verdejo, Espinilla, M., López, J.L., Melguizo, F.J.: Assessment of sustainable development objectives in smart labs: technology and sustainability at the service of society. Sustain. Cities Soc. **77**, 103559 (2022). https://doi.org/10.1016/j.scs.2021.103559
24. Zafari, F., Gkelias, A., Leung, K.K.: A survey of indoor localization systems and technologies. IEEE Commun. Surv. Tutor. **21**(3), 2568–2599 (2019)

Smart Home Interface Design: An Information Architecture Approach

Tracy Hernandez(✉), Adrian Lara, Gustavo Lopez, and Luis Quesada

Universidad de Costa Rica, Ciudad Universitaria Rodrigo Facio, San Pedro, Costa Rica
{tracy.hernandez,adrian.lara,gustavo.lopezherrera, luis.quesada}@ucr.ac.cr

Abstract. The usage of IoT devices is rapidly growing. Many users may want to add them to their homes to automate certain tasks, help themselves with information, or monitor their environment continuously. Quick IoT development is taking place on privacy, protocols, and other areas. However, the user interface area is being left out of the efforts. Mobile applications and websites are focused on technical requirements only. Research is not focused on the interaction of the user with the application, so standards and/or tendencies may not be utilized. This investigation aims to implement a smart house application interface focused on the user instead of the technical requirements. To start, the card sorting technique is used to group IoT devices into meaningful groups for users. Then, an interface prototype of a smart house application is created with the feedback obtained from the card sorting activity. Finally, the prototype is evaluated by using a standardized questionnaire.

Keywords: IoT · Card Sorting · HCI · Smart Home

1 Introduction

The Internet of Things (IoT) is a technology that allows the communication of various devices and sensors through a network to provide various solutions and advantages for users [1]. In this field, there are a wide variety of devices that different companies create and each one follows its communication protocols without direct human intervention. This entails a great challenge regarding interoperability, privacy, and security [2]. These challenges are under constant research. However, the user interaction with these devices and technologies is left behind.

This research paper aims to investigate and improve the user experience of IoT devices by employing the card sorting technique to group these devices based on the participant's perception. Subsequently, the study aims to design and develop an intuitive user interface with the ultimate goal of evaluating its usability.

In recent years, the rapid proliferation of IoT devices has introduced new challenges for users in managing and comprehending the diverse range of interconnected devices. Users often face confusion and inefficiencies due to the varying functionalities and complex user interfaces of these devices. Therefore, this

J. Bravo and G. Urzáiz (Eds.): UCAmI 2023, LNNS 841, pp. 90–101, 2023.
https://doi.org/10.1007/978-3-031-48590-9_9

research seeks to address the problem of enhancing user experience and simplifying device management in the IoT ecosystem.

Through the dissemination of findings and implications, this research also aims to support future endeavors in designing user-centric IoT interfaces and pave the way for wider adoption of IoT technologies in various domains.

The following section of this document describes the background information as well as the most important concepts regarding this research. Section 3 shows the related work and how this research is different. Section 4 focuses on the methodology that was followed during the development of this research paper. Section 5 shows the results obtained from the experiments that took place. And finally, Sect. 6 focuses on the conclusions and future work following this paper.

2 Background and Conceptual Framework

2.1 Background

There are limited research papers regarding the interaction of people with IoT systems. Most development has taken place in corporations and industries such as Amazon [3], Google [4], and Apple [5]. However, a major concern is that the applications developed may be significantly difficult to use.

There are other approaches when it comes to managing IoT devices. For example, AT&T developed a Smart Home Manager mobile application. However, its main focus is on network visibility and control [6]. Google also has an application named Home Assistant. Its downside is that it may be difficult to set up [7].

Despite that, there is a tendency to have two types of IoT systems regarding commercial products: person-centric and home-centric [8]. Table 1 and Table 2 summarize part of the findings of Lynch et all. This research paper is still valid if privacy is taken into consideration. All devices in a smart home can be managed from a single app since they are shared among the house members while using customized profiles. On the other hand, everyone is able to keep their personal data for themselves.

First, on person-centric systems, as shown in Table 1 there are fewer categories to cover. Also, the behavior of the applications is similar in all cases. The person wears or uses a device that collects specific data. Then, all the data is displayed using an application, typically a mobile one. Finally, the application may send notifications to the user. These systems tend to be very user-friendly systems

Many companies in the industry have developed their person-centric systems, in which they combine all the previous categories (or most of them) into one single application that is still usable.

For example, the application Zepp Life developed by Huami allows the person to synchronize three different types of devices: band, watch, and scale. Those devices allow the user to gather data from the three first categories shown in Table 1.

Table 1. Person-Centric IoT Systems.

Category	User Action	Interaction with User
Sleep quality, cerebral activity	Device or application usage	Alarms, notifications
Body and fitness tracking	Device or application usage, movement	Alarms, notifications, interactive objects
Weight tracking	Application, web page, scales	Alarms, notifications, interactive object
Audio-visual logging, Event logging	Device usage	Alarms, notifications, interactive object

Figure 1 is an example of what the application looks like. After synchronizing the devices, the user has access to an application that shows the data it has collected. It shows insights on changes that can be potentially beneficial for the person, and also historic data for comparison.

In summary, person-centric IoT systems combine most of the functions needed for a user in one single application easy to use.

On the other hand, there are home-centric IoT systems that cover a wider variety of categories and therefore functions as well, as shown in Table 2.

Home-centric IoT systems are more complex than person-centric systems. There are different functions that users may be performing. While there is a category division in Lynch et all's work, there is no such division in the applications.

Table 2. Home-Centric IoT Systems.

Category	User Action	Interaction with User
Home appliances	Application, Interactive device	Alarms, notifications, application
Home security	Application, Interactive device, movement	Alarms, notifications, environmental alteration
Utility measurement, Environmental awareness	Application, Interactive device	Environmental and object alteration
Garden monitoring	Plant watering	Notifications
Frequency and location tracking	Object movement, application	Application, notification
Reminders and Notification	Object usage	Application, object change
Configurable platform	Object usage, application	Application, notifications, environmental alteration

In the industry, home-centric IoT systems tend to group up as many functionalities as possible under one application. While that could seem beneficial for the user to access all of the functionalities under a single pane, it may add complexity to the application's usage as well. Also, the added complexity may outweigh the benefit of grouping up functionalities.

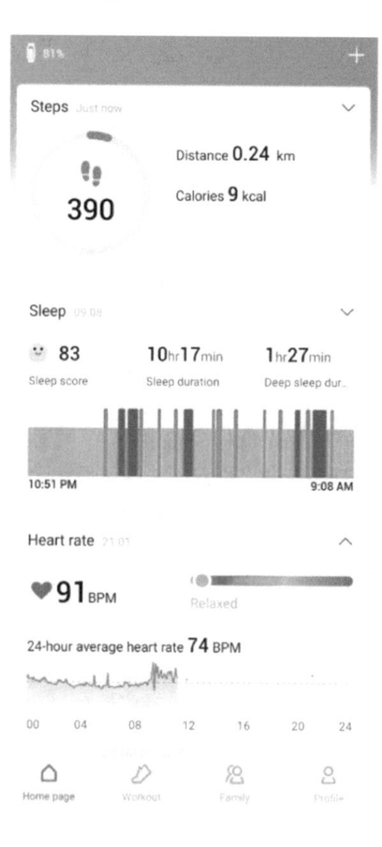

Fig. 1. Zepp Life

Companies are investing in the development of several IoT devices of their own, which can be managed from a single application. Since there are no standards or universal communication protocols yet, this scenario adds to the complexity that users face when adding IoT devices to their homes.

Some vendors or application developers try to create platforms that allow the person to manage different devices, from different vendors from the same console. However, in this case, there is also added complexity as the user is expected to know certain technical information, for example, connection protocols used in the devices, as shown in Fig. 2.

Either way, home-centric IoT systems are not user-friendly. First because of the amount of functionalities added in them. Second, because of the different devices, brands, and protocols that should not be of interest to the end user.

Fig. 2. Tuya Smart

2.2 Conceptual Framework

The term **ubiquitous computing** (also known as UbiComp) refers to a paradigm in which humans interact with several small computing devices. These devices are integrated into each aspect of daily life, without disrupting human activities [11].

On the other hand, the **Internet of Things**, or IoT for its acronym, is a term that refers to the existence of billions of devices that are collecting data about the environment in which they are located. These devices are connected to the network to send collected data to be analyzed [12].

IoT can be seen as a specialization of UbiComp, where the main difference is that in IoT there is internet connectivity between devices, which is not the case with ubicomp [13]. While the focus of this paper is not on the protocols being used, it is important to mention that the terms may be used interchangeably in the industry.

To expand further on IoT. There are three components that create what IoT is. The first one is *Things* which makes reference to adding computing capabilities to objects that are utilized in people's daily activities. The second one is *Semantic* which is related to the treatment given to the data collected by the IoT devices. The last one is *Internet* which has to do with the connectivity of the devices with a central point in which it will be treated [14]. Unfortunately, in no area of IoT, the interaction of people with these devices and systems is taken into consideration.

The meaning of **Human-Computer Interaction** (best known as HCI) is also important. While it is difficult to define what HCI means, a starting point is needed. HCI is a discipline that aims to produce a fit between the user, the device, and the services provided by the device [15]. It has to be present from the design stage to the evaluation and implementation. And finally, it also has to consider all the phenomena associated with the stages.

It is also of great importance to bring up the terms **usability** and **utility** when it comes to IoT. First, Utility according to the Cambridge Dictionary, has to do with the usefulness of a device, in this case, mostly in a practical way. On the other hand, the same source defines Usability as the degree to which it is easy to use a device, in this case [16,17].

So, in most cases, IoT systems may have utility to a certain extent. Once the device has been set up, the user can interact with it to simplify daily tasks. However, the lack of HCI involvement in the development of these systems may negatively impact the user experience. The solution should be developed by understanding the potential users and their problems [18]. So, many of these devices are frustrating for the end user at any stage from device synchronization to daily interaction.

3 Related Work

Most efforts in IoT application design have been focused on the technology side, without taking into consideration the user experience. For example, a survey shows the areas in which IoT Research has been taking place. Most of the papers cited focus on areas such as security, authentication, encryption, and protocols, among others. However, topics such as HCI or user-friendly interface designs are not mentioned [19].

However, one research paper was found that focuses on the HCI side. Fauquex et al. propose a new methodology that they call PAwEn, based on the combination of design thinking and user-centered design methodologies. The researchers take into consideration that the IoT environment is different from the traditional ones, in the sense that the devices themselves may be interacting with each other. Also, the addition of IoT technology into the day-to-day environment should not be intrusive, and therefore, new ways to interact with these systems must be created [20].

While there are certain similarities in terms of the methodology followed, there are also some key differences. Both the PAwEn methodology and this current paper follow the steps of Research, Design, Prototype, Evaluation, and Refinement in the same order. However, the ideas behind the stages are different, as shown in Table 3.

Table 3. Methodology Differences

Stage	PAwEn	This Research
Research	Literature review, questionnaires, interviews, surveys, affinity analysis	Literature review, card-sorting, surveys
Design	Ideation, scenarios, task flow	Ideation, scenarios, task flow
Prototype	High and low fidelity prototypes	High fidelity prototype
Evaluate	Heuristic evaluation, user testing	Standardized AttrakDiff questionnaire

One of the main differences is the research performed before working on the design phase. In this case, a card-sorting activity was chosen because there is a wide variety of IoT devices in the market that may be difficult to group up into categories meaningful for users if their opinion is not taken into consideration.

The second most important difference is the evaluation tool chosen. In this case, AttrakDiff was utilized as it has already been used previously in IoT environments [21]. It also sheds light on the specific areas of improvement to take into consideration based on a standard questionnaire.

4 Methodology

The main goal of this paper is to design an IoT User Interface for a home-centric system by using the card sorting technique to group up devices into categories that are meaningful for potential end users. After, the prototype is evaluated with a standardized questionnaire to assess its usability.

4.1 Card Sorting Activity

First, applications such as Tuya Smart, Amazon Alexa, and devices in the market were taken into consideration to make a list of home-centric IoT devices. The list contains 47 IoT devices that were selected taking into consideration the most common ones in already existing applications.

Then, following up with the card sorting best practices, 17 people participated in this first stage. First, one group of 9 participants and later another one with 8 participants were selected. The participants are all university students with technical backgrounds. However, they were not entirely familiar with the usage of IoT applications focused on Smart Homes. The instructions given were the following:

- Group up the cards. If there are doubts about certain cards, those can be left as *non-categorized*
- Name the groups created. The assigned name has to be meaningful.
- There are no limits in terms of the number of groups or hierarchy constraints.
- Right after the card sorting was over, a survey was conducted to gather extra feedback.

4.2 User Interface Prototype Design

The results obtained from the card sorting experiment were used to create a User Interface Prototype for a potential home-centric application. This interface has 10 categories which were the most common among the participants.

However, most participants stated that grouping up the devices was a very challenging task. So the groups as well as the feedback obtained were utilized to design a potential prototype.

Also, the *Figma* web application was used to create the prototype and the flows between screens that are to be evaluated in the following stage.

4.3 User Interface Prototype Evaluation

Once the prototype was designed and the flows were ready 20 participants were selected to evaluate it. This group of participants is different from the previous one. These participants ranged from 21 to 35 years old, where 78% stated that they are used to working with technology and the other 22% have regular experience with technology.

The prototype was presented to the team, only mentioning that not all functionalities could be tested due to it being a prototype in the early stages of design. The participants were asked to interact with the prototype and after they were done, they were given an AttrakDiff evaluation in the form of a Google Forms link to evaluate the prototype.

5 Results

5.1 Card Sorting Activity

After analyzing the data obtained from the 17 participants of the card sorting activity, 10 categories were found as shown in Table 4.

Some devices were easier to classify in comparison to others. In this case, Table 5 shows a summary of the classification difficulty perceived by the participants during the activity.

In detail, devices belonging to the categories *Kitchen, Laundry, Sensors, and Lightning* were easy to classify. Then, devices in the categories *Security, Home, and Entertainment* had a classification difficulty of medium. The rest of the categories were either hard or very hard.

It is important to note that during the experiment, several participants stated that it was confusing to classify so many devices. They also expressed having a hard time with certain devices that seemed to belong in more than just one category. However, even if they recognized the difficulty of classifying the devices, they also explained how there were certain devices they considered that should have been part of the activity voice assistants (Alexa specifically was mentioned), alarms, and biometric readers are some examples.

Table 4. Card Sorting Activity: Categories

Category	Number of Devices
Environment	4
Lightning	6
Kitchen	11
Laundry	4
Entertainment	6
Technological	2
Security	3
Home	4
Networking	2
Sensors	5

Table 5. Card Sorting Activity: Classification Difficulty

Difficulty	Number of Devices
Easy	25
Medium	13
Hard	8
Very Hard	1

5.2 User Interface Prototype Design

From the card sorting activity information gathered a prototype was designed. In this case, the participants had difficulties classifying the devices, but at the same time, many would have added other devices. This feedback was crucial to design the prototype to be tested.

The prototype design suggests the usage of automatic or simpler ways for a potential user to add IoT devices to the app for management. In this case, an *Automatic Mode* is added in which the application could use different technologies to find the devices such as Bluetooth, and WiFi, among others. There was also a *Manual Mode* added in case the user knows which protocol is being used and wants to select it right away or in case the vendor uses QR codes in the packaging to scan and add the device into the app. Finally, inside the manual mode, the option of following all the steps manually, as done with other apps, is kept for those users who prefer this method. All of these options are shown in Fig. 3.

In Fig. 3, it is shown in the *Home Page*, that the user is given a summary of the number of devices being managed, turned on, or unreachable or failed to connect. Also, they have a section for custom categories that are more meaningful and completely customizable. For example, to group up the devices that are used in the kitchen in a *room* named the same way. Finally, at the bottom, the default

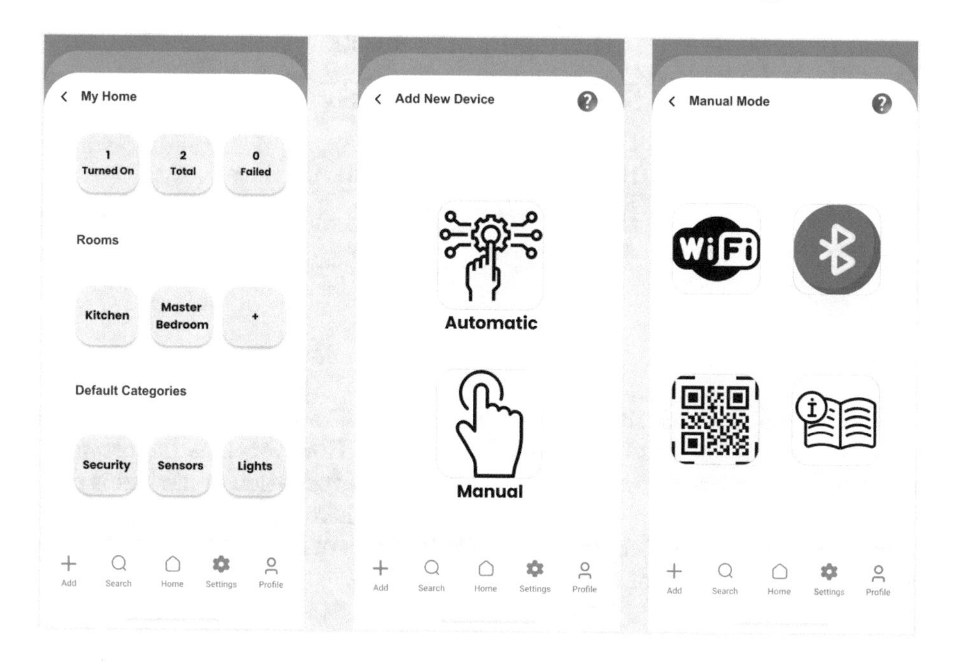

Fig. 3. Prototype Examples

categories, obtained from the card-sorting activity are added, as a second way to browse through the devices.

The main goal which this design is to provide the user with an automatic way to add devices to simplify the usage of the app. Also, allow customization in terms of rooms that are meaningful. Finally, some people are already used to the way in which IoT Applications work at the moment and rather a way in which they can browse through all of the options available.

5.3 User Interface Prototype Evaluation

To evaluate the prototype described in the previous section, an AttrakDiff standard questionnaire was utilized. It is possible to evaluate the usability of a prototype with this type of standardized evaluation with a group of 22 people, which is the reason why it was chosen [21].

While there is room for improvement in the prototype, the results show that the participants have a positive tendency regarding the prototype as shown in Table 6. The first row refers to the usability of the product. The second one indicates how identified the user feels with the system. The third is about the stimulation that the system generates in the user. And the last one refers to the overall attractiveness. All rows work on a scale from 1 to 7 [22].

From the feedback gathered, 85% of the participants stated that they would prefer an automatic way to find IoT devices instead of a manual procedure. According to them, this seems to be the aspect of the prototype that positively

Table 6. AttrakDiff Results

Factor	Score
Pragmatic Quality	4.2
Hedonic Quality - Iden	4.3
Hedonic Quality - Stim	3.8
Attractiveness	3.9

impacted them and made them feel like it was an innovative solution compared to others.

Finally, the feedback that seems to have drawn the AttrakDiff results closer to a middle point is mostly on the aesthetic side. However, it is important to mention that some participants stated that the categorization of devices is difficult, and while some wanted more categories and others wanted fewer categories, most of them agreed on liking the flexibility in terms of being able to create their own categories in terms of house rooms.

6 Conclusions

The inclusion of HCI in the design and development of IoT applications for Smart Homes could be very beneficial. The card-sorting activity shed light on the difficulty that categorizing a wide variety of devices can carry. So, looking for alternatives to add and manage IoT devices from mobile applications is highly valuable for end users. Actually, the idea of automation and flexibility was appreciated by the users even where there were possible improvements.

On the other hand, it is crucial to consider if it is really needed to add a great amount of IoT devices and categories, which in all stages of the research was a pain point for the participants. Flexibility in this aspect is of great importance, but mostly avoiding confusion for the users should be a must.

The IoT devices demand will only keep increasing, which means that more efforts should be put into applications that are designed to facilitate these technological tasks for the end users.

Finally, there is still room for improvement and research. In future work, a prototype like the one presented in this research paper can be expanded to be tested more in-depth. Also, the attractiveness of the prototype must be enhanced as well. To conclude, there are more activities besides the card-sorting one, that could be employed to shed light on more aspects in which these efforts can be further enhanced.

References

1. Kumar, S., Tiwari, P., Zymbler, M.: Internet of things is a revolutionary approach for future technology enhancement: a review. J. Big Data **6**(1), 1–21 (2019). https://doi.org/10.1186/s40537-019-0268-2
2. Karthik, V., Krishna, R., Sanyasi Rao, A.: A study on IoT technologies, standards and protocols (2021). https://doi.org/10.17697/ibmrd/2021/v10i2/166798
3. Amazon. Alexa Skills kit. https://developer.amazon.com/en-US/alexa/alexa-skills-kit
4. Google. What a Great User Experience Looks Like. https://developer.android.com/quality/user-experience
5. Apple. Designing for iOS. https://developer.apple.com/design/human-interface-guidelines/designing-for-ios
6. AT&T. AT&T Smart Home Manager. https://www.att.com/internet/smart-home/
7. Google. Google Assistant. https://www.home-assistant.io/integrations/google_assistant/
8. Lynch Koreshoff, T., Robertson, T., Leong, T.W.: Internet of things: a review of literature and products (2013). https://doi.org/10.1145/2541016.2541048
9. Zepp Life, Huami. https://play.google.com/store/apps/details?id=com.xiaomi.hm.health&hl=es&gl=US
10. Tuya Smart, Tuya. https://www.tuya.com/solution/scene/smart-home
11. Guettala, M., Bourekkache, S., Kazar, O.: Ubiquitous learning a new challenge of ubiquitous computing: state of the art (2021). https://doi.org/10.1109/ICISAT54145.2021.9678434
12. Patel, K.K., Patel, S.M., Scholar, P.: Internet of Things-IOT: Definition, Characteristics, Architecture, Enabling Technologies, Application & Future Challenges (2016)
13. Andrade, R.M.C., Carvalho, R.M., de Araújo, I.L., Oliveira, K.M., Maia, M.E.F.: What changes from ubiquitous computing to internet of things in interaction evaluation? In: Streitz, N., Markopoulos, P. (eds.) DAPI 2017. LNCS, vol. 10291, pp. 3–21. Springer, Cham (2017). https://doi.org/10.1007/978-3-319-58697-7_1
14. Corcoran, P.: The internet of things: why now, and what's next? (2016). https://doi.org/10.1109/MCE.2015.2484659
15. Karray, F., Alemzadeh, M., Saleh, J., Arab, M.: Human-computer interaction: overview on state of the art (2018). https://doi.org/10.21307/ijssis-2017-283
16. Usability. In: Cambridge Dictionary (2023). https://dictionary.cambridge.org/dictionary/english/usability
17. Utility. In: Cambridge Dictionary (2023). https://dictionary.cambridge.org/dictionary/english/utility
18. Rowland, C., Goodman, E., Charlier, M., Light, A., Lui, A.: Designing Connected Products. O'Reilly Media, Sebastopol (2015)
19. Mohamad, M., Haslina, W.: Current research on internet of things (IoT) security: a survey (2019). https://doi.org/10.1016/j.comnet.2018.11.025
20. Fauquex, M., Goyal, S., Evequoz, F., Bocchi, Y.: Creating people-aware IoT applications by combining design thinking and user-centered design methods (2015). https://doi.org/10.1109/WF-IoT.2015.7389027
21. Diaz-Orozco, I., Lopez, G., Quesada, L., Guerrero, L.: UX Evaluation with standardized questionnaires in ubiquitous computing and ambient intelligence: a systematic literature review (2021). https://doi.org/10.1155/2021/5518722
22. AttrakDiff. https://www.attrakdiff.de/sience-en.html#messen

Evaluation of the Jetson Nano on the Analysis of Risk Patterns in Multimodal Traffic Intersections

Josu Gomez, Hugo Landaluce(✉), and Ignacio Angulo

Faculty of Engineering, University of Deusto, Avda. Universidades, 24, 48007 Bilbao, Spain
jogoar@opendeusto.es, {hlandaluce,ignacio.angulo}@deusto.es

Abstract. The analysis of risk patterns in multimodal intersections is fundamental to ensure safety in cities. Government policies regarding data protection aim to guarantee the privacy of individuals, but they complicate the timely deployment of traffic cameras on the roads for these purposes. Traditionally, this requires municipal authorities to have personnel visually inspect high-conflict traffic intersections. As such, a system capable of determining the risk patterns of different areas using embedded devices at the edge without the need of storing or transmitting video is proposed, ensuring compliance with data protection regulations and improving latency without network delays for processing. Experiments that focus on the relevant performance, energy consumption and accuracy metrics of the system with the constrained hardware of the Jetson Nano show real time inference speeds with respect to a non-constrained system on the analysis of the different factors to evaluate the risk of a traffic intersection under a low energy consumption scenario, and the room for increased power consumption to performance ratio of higher end accelerated hardware is showcased, depending on the power profile desired, without sacrificing accuracy in such an ITS context.

Keywords: Internet of Things (IoT) · Edge Computing · ITS · risk factors

1 Introduction

The increasing demand for safe and connected cities has highlighted the importance of Intelligent Transportation Systems (ITSs) capable of analyzing risk areas in multimodal intersections where not only vehicles, but pedestrians are also present, given that 40–50% of road crashes in most countries are caused in traffic intersections [1], due to a varied array of factors such as the ones shown in Fig. 1. To address privacy concerns, specialized timely deployment of cameras for surveillance becomes complicated under government data protection policies. Consequently, many municipal authorities still rely on error prone visual inspections by traffic technician personnel to assess high-conflict traffic intersections. The accuracy achieved by such inspections often varies on the effort invested, time of desired delivery or fatigue of the staff.

This work has been part-funded by the Basque Government under the project PORTAERA (ZL-2023/00148)

J. Bravo and G. Urzáiz (Eds.): UCAmI 2023, LNNS 841, pp. 102–113, 2023.
https://doi.org/10.1007/978-3-031-48590-9_10

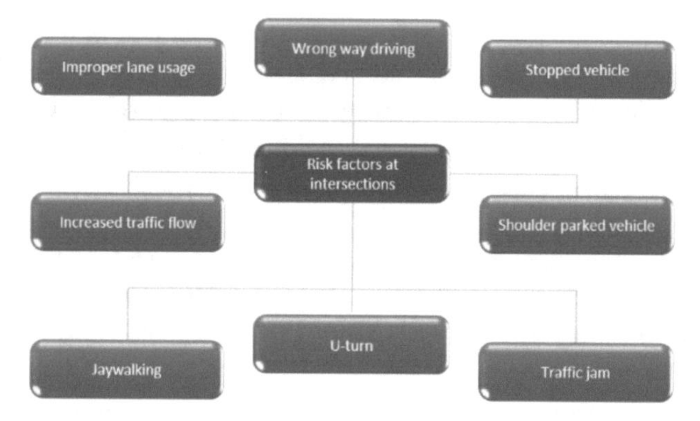

Fig. 1. Collection of various factors affecting the risk of a traffic intersection.

The advancements in computer vision and Artificial Intelligence (AI) have revolutionized video image recognition and tracking, enabling a wide range of applications in various industries [2]. One approach of attention is the utilization of edge computing for these tasks [3]. Edge computing refers to the practice of locally processing data closer to the source, reducing cost, energy consumption, data processing latency, network bandwidth usage requirements and dependency on the cloud.

This paper proposes a solution to automate the risk assessment process using an edge computing embedded system that leverages the Jetson Nano. By adopting this approach, compliance with data protection regulations that forbid transmitting sensitive video to the cloud can be ensured while still maintaining the effectiveness of automatic risk assessment. Furthermore, minimizing energy consumption is a critical consideration for the environment, combated using autonomous embedded devices [4]. Thus, the paper includes an evaluation of the lower end Jetson Nano device that serves as a baseline of hardware accelerated edge devices, specifically focusing on its performance, energy consumption and accuracy. The evaluation explores the impact of different batch sizes, quantization precisions and power modes on the used models, TrafficCamNet, VehyleTypeNet and VehicleMakeNet, with the goal to demonstrate that the low end Nvidia Jetson Nano, despite their constrained hardware, can achieve a notable level of performance and energy consumption tradeoff balance in analyzing risk factors compared to a non-constrained system.

2 Background

Edge hardware accelerators are increasingly being used as the main on-edge AI model execution units because of their low power consumption and high performance. Several hardware devices and software solutions to implement ITS computer vision systems are presented.

2.1 Hardware Acceleration on the Edge

At large, three types of on-edge hardware accelerators can be found in the literature.

Graphic Processing Unit (GPU) Accelerators. They show good results under general-purpose applications [5]. Some of the most popular hardware accelerators fall into this category, like the one chosen for the experimentation, the Jetson Nano [6]. They facilitate massive parallel computation commonly encountered in AI workloads.

Application Specific Integrated Circuit (ASIC) Accelerators. Having suffered a massive rise in popularity, they can take advantage of more precise strategies to accelerate calculations. Examples include the Tensor Processing Unit (TPU) from Google in the form of Coral USB accelerators, the Deep Learning Accelerator (DLA) present in many Nvidia Jetson devices or Intel's Vision Processing Unit (VPI) of the Movidius neural compute stick.

Field Programmable Gate Array (FPGA) Accelerators. They are cost and energy efficient solutions that require little adaptation to the hardware thanks to their reconfigurable nature.

Regarding the rest of the acceleration technologies, GPUs represent the most widely adopted solution, and therefore, currently offer greater versatility as they have more software support, ideal for the development of parallel performant applications.

2.2 Software Solutions

In the realm of ITS, video image recognition and object tracking are vital components for enhancing road safety and optimizing traffic management. There exist several efforts focused on applying deep learning methods for solving various traffic related problems. In [7] and [8] deep CNNs are used to recognize traffic signals and fire respectively. Moreover [9] deploys a real-time face recognition on several edge platforms providing details about the executing time, energy consumption and memory usage on those platforms. You Only Look Once (YOLO) [10], an open-source system aimed at detecting objects in real-time, has also been widely used to detect objects in images such as [11] for vehicle detection and [12] for counting vehicles.

While object detection is a necessary step in the scene understanding, objects must be related to one another so that there is the ability to discern them. In the efforts of matching a single detection with another, multiple algorithms have been considered, going from the simple IOU tracker [13], which compares the region of overlap between bounding boxes to a certain threshold, to using a more advanced Kalman filter [14] component that allows future behavior prediction or ending in discriminative correlation filter [15] visual tracking that adjusts the correlation filter dynamically. Each has their own performance characteristic that must be adjusted to the problem in question.

Other suitable for ITS are Google Cloud Vision API - Occupancy Analytics Model [16] and Microsoft Azure Cognitive Services - Spatial Analysis Operations [17]. They provide a suite of AI-powered tools that enable the analysis of video streams to extract valuable information such as object detection, tracking and spatial relationships. Notice these solutions primarily operate in a cloud-based environment and may not be suitable for edge computing scenarios with constrained hardware resources.

3 System Design

To be able to take advantage of the performance of the used hardware, Nvidia GPU accelerated related components are predominantly used. The Nvidia Jetson, as mentioned, are small but powerful computers that can execute otherwise costly neural networks thanks to their onboard GPU and fixed function processing accelerator inclusions. The Jetson Nano and its Jetson Xavier NX comparative counterpart were specifically chosen as a confident minimum baseline of hardware accelerated edge computing as they stand on the lower range of the Nvidia Jetson family.

The rest of the software was executed on top of the Ubuntu based Jetson Linux operating system, specifically tuned for their ARM based Tegra System on a Chip (SoC). In addition, the GStreamer based GPU accelerated DeepStream video processing toolkit was a key component of the integration of the pipeline components because it allowed to harness the capabilities of all the accelerated processing components. With that, Nvidia TAO Toolkit's models were deployed using the neural network execution engine TensorRT.

3.1 Software Architecture

Using the corresponding tools, the software architecture is organized into multiple individual elements connected to one another (see Fig. 2). Each one performs a critical video processing task in the complete operation of the pipeline in its entirety, being able to utilize the advanced capabilities of video processing and hardware access that edge computing embedded devices offer.

Fig. 2. Software architecture organization into multiple elements.

Video Input and Decoding. For ingesting and subsequently processing video, it is essential to extract the data the video stream contains by understanding each individual frame so that analysis can be performed on them. Most surveillance cameras use the H.264 codecs [18] to transfer video. This way, it is not feasible to work directly with the encoded footage due to it not directly representing the real data. Here is where video decoders come into play, which decode the frames so that they can be worked with by transforming them into raw data consisting of a certain agreed upon format. Unfortunately, decoding is a costly process, mainly when working with multiple video streams as is the case, and for that, the advanced hardware accelerated decoding speeds up the process by taking advantage of the computing power of the graphics card as well. The resulting increase in throughput enables more processing to happen at the same time.

Pre-processing. The higher throughput edge devices can give due to not being constrained by network limitations and being able to ingest large quantities of video feeds must be accompanied by the appropriate video processing requirements that would not tank the performance of the system. Optimally, the full system resources in the form of parallel processing instead of simply increasing the raw computing power of the machine itself would be used. Unlike conventional threading, batch processing is the more adequate solution when using GPU enabled devices since it exploits the inherent parallelism of the massively multicore systems that are the GPUs by performing the same operation at the same time to several different inputs, that is, treating individual values as whole [19].

Primary and Secondary Inference. To understand the context of the scenario, the initial inference work that must be carried out consists of detecting the objects of interest, vehicles and pedestrians, mainly. This is done using the purpose built primary multi-class object detection model TrafficCamNet, that can regress the location and dimensions of the classes. Inspired by YOLO, the model uses the DetectNet_v2 architecture to regress the location and dimensions of the bounding boxes in an image grid in the form of four parameters for each cell. An 18-layer deep ResNet [20] is used as a feature extractor to counter the problem of vanishing/exploding gradients [21], with a Non-Maximal Suppression (NMS) for processing boxes. Besides the primary detector, varied object characteristics can be extracted using secondary class specific inference to distinguish vehicle models with VehicleTypeNet and make with VehicleMakeNet.

Tracking. Live video, having the additional time component not present in static images, requires the use of target re-association between frames in order to correctly identify and trace the movement patterns required for further analysis. In a traffic intersection context, multiple objects must be tracked simultaneously and, ideally, the tracker would be resilient of occlusion for when tracks are inevitably partially occluded by the environment, such as trees or traffic lights. Taking that into account, he discriminative correlation filter based multi object tracker, NvDCF, is used, which tracks objects based on location, bounding box size and visual appearance similarity.

Analysis. Context specific scene annotations are required to perform the risk analysis of the scene that define the corresponding areas to account for, as shown in Fig. 3, allowing a detailed object, vehicle and pedestrian specifically, trajectory processing in the form of filtering tracks inside an area of interest, counting occurrences inside a certain region, detecting the direction of flow, signaling overcrowding and detecting line crossings. Line crossing detection is performed by finding the possible line segment intersection point as described in [22]. Area related operations are performed using a standard point in polygon algorithm, the even-odd rule algorithm that counts how many times a line starting from the point in question intersects the polygon, as outlined by [23]. If the point is inside the intersection count will be odd and if it is outside, even, proven by Jordan curve theorem [24].

Distribution. Finally, the collected data must be transmitted to the corresponding consumer entities in a timely fashion so that they can act on it. With the need of arbitrary data in the form of the context information containing the analytics results has to be broadcasted, asynchronous communication comes into play, in which, unlike synchronous communication protocols like the prevalent HTTP, the producer of the messages does not need to be aware of its consumers, freeing it from having to wait until the message is received.

Fig. 3. Multiple simple traffic intersection annotation configurations shown as colored lines. Possible uses from left-to-right, top-to-bottom: jaywalking detection by detecting pedestrians inside the region of interest (yellow), pedestrian flow counting by understanding the directions pedestrians cross the crosswalk (yellow/green), traffic overcrowding detection by counting the number of stopped vehicles at the traffic light (orange) and wrong way driving detection by checking which vehicles are turning incorrectly (green). (Color figure online)

4 Experiments and Results

The experiments aim to measure the multiple performance characteristics of a low-end low resource usage hardware accelerated embedded edge device in comparison to a more powerful alternative and non-constrained hardware. For that, the Jetson Nano was chosen, alongside the slightly more powerful Jetson Xavier NX and a GeForce GTX 1650 (Mobile) GPU. The following Table 1 gives a summary of the main compute specifications of each device.

Table 1. Device specification comparison.

Device	CPU cores	GPU cores	Tensor cores	GPU memory	Max power
Jetson Nano	4	128	0	4 GB (shared)	10 W
Jetson Xavier NX	6	384	48	8 GB (shared)	20 W
GTX 1650 (Mobile)	N/A	896	0	4 GB	30 W

All experiments have been carried out by using the maximum power modes, fixing GPU clocks and with the most performant quantization precision unless stated otherwise, quantization being the reduction of precision of the data types used by the network [25], usually resulting in a reduction of accuracy but improving the performance and memory. The Jetson Nano supports both 32-bit and 16-bit floating point quantization and the Jetson Xavier NX allows for 8-bit integer precision as well. Batch processing combats performance and energy consumption degradations caused by variable number of streams and targets. The key indicators chosen for the comparison are performance, energy consumption and accuracy.

4.1 Performance

Performance of the inference using the TrafficCamNet, VehicleTypeNet and VehicleMakeNet was measured in throughput or Frames Per Second (FPS), that is, the number of inferences processed in a second, using the trtexec tool averaged for 1000 inferences. For a more accurate measurement improving GPU utilization, the host to device and device to host data transfers were disabled since, once batched, frames already live on the GPU.

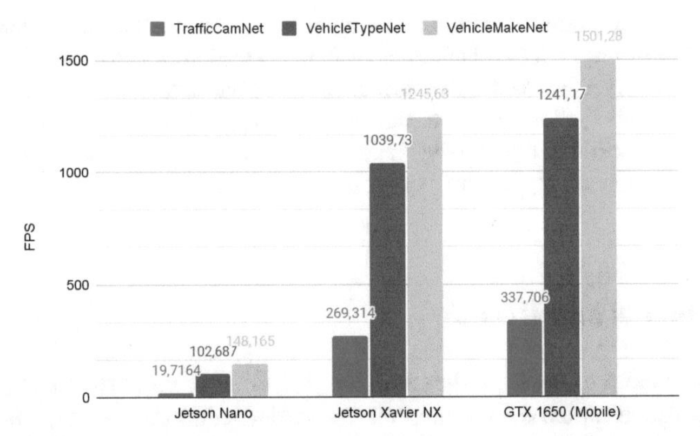

Fig. 4. FPS comparison over a batch size of 1. INT8 precision is used in the Jetson Xavier NX and the GTX 1650 (Mobile) and FP16 in the Jetson Nano.

Figure 4 metrics show real time inference performance of the Jetson Nano, while not quite reaching the 30 FPS mark, delivering an acceptable 20 FPS. It is relevant to note that the highly increased performance of the Jetson NX, although it falls behind the GTX 1650 (Mobile), it highlights the large room for inference gains thanks to the use of 8-bit integer quantization precision.

The capabilities of simultaneous processing of multiple streams by increasing the batch size is also shown in Fig. 5 with a theoretical around 128 times throughput gains limited by the memory of 4 GB of the device using the primary model. In practice, inference is just part of the system as a whole and GPU memory is shared with other running programs in the case of the Nvidia Jetson family and subsequent processing operations might not be able to keep up with the FPSs.

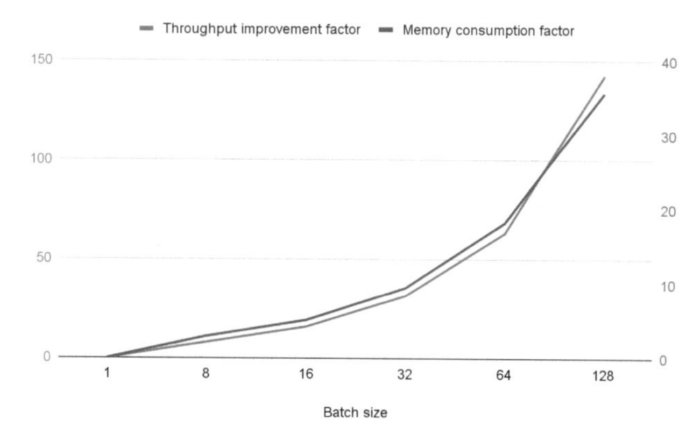

Fig. 5. Factor of throughput improvement and memory consumption of batch sizes of the Jetson Nano that shows the similar trend of throughput improvement and memory consumption.

4.2 Energy Consumption

Maximal expected energy consumption of the Nvidia Jetson devices was inspected using the Hameg HM8115–2 power-meter for more accurate readings than with software-based tools. Power drain for Fig. 6 was measured over time at idle and during the previous maximal inference workloads for an hour to simulate peak driving times in different power profiles to correctly assess the power to performance ratio, also showing how stable the devices can be, not experiencing any noticeable power spikes.

In a real-world scenario, such as the one indicated in the following section, power consumption can vary slightly, mainly in the power profiles with more budget, due to the small changes seen in CPU usage percentage caused by changes in the number of objects of interest in each frame. However, the bulk of the workload is carried out by the GPU which continuously operates at full no matter the situation.

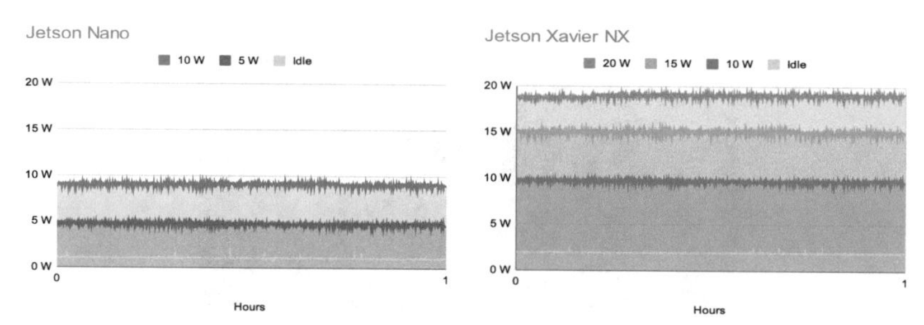

Fig. 6. Power consumption comparison of the different power modes of the Jetson Nano and Jetson Xavier NX over an hour of load.

Findings from Table 2 show that the Jetson Nano is best in low energy consumption while still being able to process around 15 FPS, not taking batch size into account, at its lowest power profile (5W). Additionally, a greater than 6 times power to performance ratio increase with the highest performing profile can be achieved using the comparative higher end embedded edge computing device Jetson Xavier NX.

Table 2. Performance impact of power profiles with TrafficCamNet.

Device	Power profile	GPU compute time	FPS
Jetson Nano	5W	70.7594 ms	14.0998
Jetson Nano	10 W	50.4478 ms	19.7164
Jetson Xavier NX	10 W	4.89624 ms	203.666
Jetson Xavier NX	15 W	3.82495 ms	207.656
Jetson Xavier NX	20 W	3.68555 ms	269.314

4.3 Accuracy

The accuracy of the system was measured in terms of the object detection and tracking. Unfortunately, not many datasets exist that capture the multimodal flow of pedestrians and vehicles of a traffic intersection in the perspective of a traffic camera, however, one of those few datasets made with the purpose of evaluating the accuracy of object trackers in urban mixed traffic is the Urban Tracker dataset [26].

More specifically, the Sherbrooke track filmed at the Sherbrooke/Amherst intersection in Montreal was chosen, as it is the most akin to a traffic camera perspective and can result in further challenge due to the multiple occlusions between road users. The accuracy was measured for the entire duration of the 1001 frames of the dataset. Figure 7 shows an example frame of the intersection that contains multiple objects at the same time.

Fig. 7. Sherbrooke track sample from the Urban Tracker dataset [26].

The main metrics that were considered were Multi-Object Tracking Accuracy (MOTA) and Multiple Object Tracking Precision (MOTP). MOTA measures the quality of the detection and tracking output by taking false positive, false negative and tracking id switches into account. False positives (FP) dictate when the algorithm detects an object not present in the ground truth (GT), false negatives (FN) dictate when the object is present in the ground truth but is not detected and id switches (IDS) when an object's id is incorrectly related to another.

$$MOTA = 1 - \frac{|FP| + |FN| + |IDS|}{|GT|} \quad (1)$$

MOTP is more closely related to exclusively the object detection in that it determines how well the bounding box of the object matches the ground truth with the true positives (TP). The association threshold (S) was an absolute distance of 90 pixels.

$$MOTP = \frac{1}{|TP|} \sum_{TP} S \quad (2)$$

As outlined in Table 3, the system can perform accurately in a traffic intersection scenario without the need of sacrificing power or performance at the edge. It is relevant to note that many bounding boxes where non-intuitively cropped towards the end of their tracks throughout the dataset and that accuracy could be further improved by using a camera angle that can consistently view the front or rear of vehicles instead of the sides in some cases.

Table 3. Accuracy evaluation results for car and pedestrian classes.

Video	Type	MOTA	MOTP
Sherbrooke	Cars	0.7719	7.14 px
	Pedestrians	0.8014	9.21 px
	All objects	0.7971	x

5 Conclusions and Future Work

This paper presents a system capable analyzing the risk patterns of multimodal intersections on the Jetson Nano using context specific annotations, evaluating its performance, energy consumption and accuracy guarantees with the purpose of determining the use of such a lower end hardware accelerated device on the edge.

Final results show the notable real time inference performance of the Jetson Nano with respect to a non-constrained system, having the lowest power consumption requirements out of all the compared devices with a power drain profile of 5 W. Capabilities of simultaneous processing of multiple streams are highlighted that result in a throughput improvement theoretically up to 128 times using the primary model limited by memory, as well as a massive room for inference gains with a greater than 6 times higher power to performance with the highest performing power profile using the comparative higher end embedded edge computing device without sacrificing inference accuracy by achieving respectable evaluation results in an ITS context.

Future efforts in the form of a more robust approach of detecting vehicles and pedestrians may result in an improved reliability for real world applications, possibly achieved by training the model in higher amounts of different traffic camera perspective data or using context specific data in domain adaption training. Likewise, incorrect recounts of objects could be further avoided by improving the visual tracking when dealing with suboptimal camera placements caused by occlusions, with the purpose of allowing a higher reuse of already placed traffic cameras that might not be able to be adjusted. Mostly, augmenting the user experience aspect when defining context aware annotations would be highly beneficial to make the process more straight forward.

References

1. Lefèvre, S., Laugier, C., Ibañez-Guzmán, J.: Risk assessment at road intersections: Comparing intention and expectation. In: 2012 IEEE Intelligent Vehicles Symposium, Madrid, Spain, pp. 165–171 (2012)
2. Yilmaz, A., Javed, O., Shah, M.: Object tracking: a survey. ACM Comput. Surv. **38**(4), 13–es (2006). https://doi.org/10.1145/1177352.1177355
3. Cao, K., Liu, Y., Meng, G., Sun, Q.: An overview on edge computing research. IEEE Access **8**, 85714–85728 (2020)
4. Mocnej, J., Miškuf, M., Papcun, P., Zolotová, I.: Impact of edge computing paradigm on energy consumption in IoT. IFAC-PapersOnLine **51**(6), 162–167 (2018)
5. Suo, J., Zhang, X., Zhang, S., Zhou, W., Shi, W.: "Feasibility Analysis of Machine Learning Optimization on GPU-based Low-cost Edges. In: 2021 IEEE SmartWorld, Ubiquitous Intelligence & Computing, Advanced & Trusted Computing, Scalable Computing & Communications, Internet of People and Smart City Innovation, Atlanta, GA, USA, pp. 89–96 (2021)
6. Zhu, J., Feng, H., Zhong, S., Yuan, T.: Performance analysis of real-time object detection on Jetson device. In: 2022 IEEE/ACIS 22nd International Conference on Computer and In-formation Science (ICIS), Zhuhai, China, pp. 156–161 (2022)
7. Vaidya, B., Paunwala, C.: Traffic sign recognition using color and spatial transformer network on GPU embedded development board. In: Nain, N., Vipparthi, S.K., Raman, B. (eds.) CVIP 2019. CCIS, vol. 1147, pp. 82–93. Springer, Singapore (2020). https://doi.org/10.1007/978-981-15-4015-8_8
8. Gotthans, J., Gotthans, T., Marsalek, R.: Deep convolutional neural network for fire detection. In: International Conference Radioelektronika, pp. 0–5 (2020)
9. Koubaa, A., Ammar, A., Kanhouch, A., AlHabashi, Y.: Cloud versus edge deployment strategies of real-time face recognition inference. IEEE Trans. Netw. Sci. Eng. **9**(1), 143–160 (2022)
10. Redmon, J., Divvala, S., Girshick, R., Farhadi, A.: You only look once: unified, real-time object detection. In: Conference Computer Vision and Pattern Recognition, vol. 2016-Decem, pp. 779–788 (2016)
11. Lin, J., Sun, M.: A YOLO-based traffic counting system. In: Conference on Technologies and Application of Artificial Intelligence, pp. 82–85 (2018)
12. Asha, C., Narasimhadhan, A.: Vehicle counting for traffic management system using YOLO and correlation filter. In: IEEE International Conference on Electronics Computing and Communication Technologies, pp. 1–6 (2018)
13. Bochinski, E.. Eiselein, V. Sikora, T.: High-Speed tracking-by-detection without using image information. In: 14th IEEE International Conference on Advanced Video and Signal Based Surveillance (AVSS), Lecce, Italy, 2017, pp. 1–6 (2017)

14. Anastasiou, A., Makrigiorgis, R., Kolios, P., Panayiotou, C.: Hyperion: a robust drone-based target tracking system. In: 2021 International Conference on Unmanned Aircraft Systems (ICUAS), pp. 927–933 (2021)
15. Farkhodov, K., Lee, S.H., Kwon, K.R.: Object Tracking using CSRT Tracker and RCNN. In: BIOIMAGING, pp. 209–212 (2020)
16. Vaithiyanathan, D. Muniraj, M.: Cloud based text extraction using google cloud vison for visually impaired applications. In: 2019 11th International Conference on Advanced Computing (ICoAC), pp. 90–96 (2019)
17. Del Sole, A. Introducing Microsoft Cognitive Services. Microsoft Computer Vision APIs Distilled: Getting Started with Cognitive Services, 1–4 (2018). https://doi.org/10.1007/978-1-4842-3342-9_1
18. Kalva, H.: The H.264 Video Coding Standard. IEEE Multimedia **13**(4), 86–90 (2006)
19. Costa, L.B., Al-Kiswany, S., Ripeanu, M.: GPU support for batch oriented workloads. In: 2009 IEEE 28th International Performance Computing and Communications Conference, Scottsdale, AZ, USA, pp. 231–238 (2009)
20. He, K., Zhang, X., Ren, S., Sun, J.: Deep Residual Learning for Image Recognition. In: 2016 IEEE Conference on Computer Vision and Pattern Recognition (CVPR), Las Vegas, NV, USA, pp. 770–778 (2016)
21. Bengio, Y., Simard, P., Frasconi, P.: Learning long-term dependencies with gradient de-scent is difficult. IEEE Trans. Neural Netw. **5**(2), 157–166 (1994)
22. Antonio, F.: Chapter IV.6: faster line segment intersection. In: Kirk, D. (ed.) Graphics Gems III, pp. 199–202. Inc, Academic Press (1992)
23. Shimrat, M.: Algorithm 112: Position of point relative to polygon. Commun. ACM **5**(8), 434 (1962)
24. Hales, T.C.: The Jordan Curve Theorem, formally and informally. Am. Math. Mon. **114**(10), 882–894 (2007)
25. Goel, A. Lu, Y., Thiruvathukal, G.K.:A Survey of methods for low-power deep learning and computer vision. In: 2020 IEEE 6th World Forum on Internet of Things (WF-IoT), New Orleans, LA, USA, pp. 1–6 (2020)
26. Jodoin, J.-P., Bilodeau, G.-A., Saunier, N.: Urban Tracker: multiple object tracking in urban mixed traffic. In: 2014 IEEE Winter Conference on Applications of Computer Vision (WACV14), Steamboat Springs, Colorado, USA, pp 885–892 (2014)

Extending LoRaWAN with Real-Time Scheduling

Ousmane Dieng[1(✉)], Rodrigo Santos[2,3], and Daniel Mosse[1]

[1] Department of Computer Science, University of Pittsburgh, Pittsburgh, USA
oud5@pitt.edu

[2] Dep. Ing. Eléctrica y de Computadoras, Universidad Nacional del Sur, Bahia Blanca, Argentina

[3] ICIC, UNS-CONICET, Bahia Blanca, Argentina

Abstract. LoRaWAN has recently become the LPWAN (low-power wide area network) standard for the Internet of Things (IoT) because of its long-distance communication reach with low-power transmissions in a single-hop system architecture, the possibility of deploying private networks, and its use of license-free ISM frequency bands (reducing deployment and operating costs). For its adoption in industry, LoRaWAN must deliver more scalable and reliable real-time data transmission. However, these characteristics are difficult to achieve due to collisions introduced by certain inherent features of the LoRaWAN network that lead to packet collisions, namely: (i) transmissions without carrier detection or collision avoidance, and (ii) retransmission mechanisms. To overcome these problems, we add a real-time, collision-free scheduling algorithm, based on graph coloring and fully compatible with LoRaWAN. We evaluate our solution through simulations on NS-3, where we consider scalability, with multiple gateways and end devices. The results show that, with several LoRaWAN parameter configurations, packet loss can be drastically reduced (to almost zero). Furthermore, With our protocol, end devices and network server transmit/receive packets within deadlines.

Keywords: LoRaWAN · Real-Time Scheduling · Graph coloring

1 Introduction

Smart-environments require a perception layer capable of providing the necessary data to process them and transform them into information and knowledge. For this, the Internet of Things (IoT) paradigm is the natural option. However, access to the different sensing and actuating nodes is not easy when they cover a very wide area. LoRa (Long Range) and in particular LoRaWAN is one of the favorite technological solutions for such deployments [4,9,12]. Examples of applications include large industrial and agricultural/environmental monitoring, given LoRa's ability to transmit to large distances at very low power.

LoRaWAN operates in the unlicensed radio-electric spectrum, where End-Devices (ED) can transmit only with a reduced duty-cycle, commonly less 1%

J. Bravo and G. Urzáiz (Eds.): UCAmI 2023, LNNS 841, pp. 114–126, 2023.
https://doi.org/10.1007/978-3-031-48590-9_11

and around 2–10 packets per day (sufficient for some applications). Using LoRa wireless technology as the physical layer, EDs exchange messages with Gateways (Gw), which are responsible for uploading the messages to the network server (NS).

LoRaWAN operates with an aloha-like protocol, where interference and collisions cause packet losses. LoRa has six orthogonal spreading factors (SFs) and the Gw may operate in 16 orthogonal channels, yielding 96 orthogonal possibilities. Increasing the SF increases both the transmission range (using the same transmission power) and the transmission time. In industrial applications, the SF is usually between 7 and 9, while larger areas (e.g., agriculture), the SFs used are between 9 and 12 to ensure that the gateways receive the messages.

As the spreading factor selected increases, the probability of producing collisions at distant gateways is also increased, as for transmitting the same message much more time is needed (higher channel occupancy) and larger coverage area.

When messages have deadlines, it is said that they are real-time [17], and network scheduling becomes mandatory to regulate access to the medium. In addition to this, periodic messages and transmission times are subject to duty-cycle constraints. Altogether, the scheduling problem has exponential characteristics, and some heuristics should be applied to order the transmission of messages for the EDs. In scenarios where Gws are close to each other, several Gws may receive the same message, providing a backup in case of gateway failure. In this way, two problems are to be solved: i) the node-gateway allocation, and ii) the time scheduling of messages subject to the different real-time and physical protocols constraints.

We expand on our initial LoRaWAN work [8,13] contributing a graph coloring technique based on the SF and ED-Gw coverage to provide a better feasible real-time schedule. Experiments based on the NS-3 network simulator tool are presented. The packet loss percentage (also known as packet loss rate) are shown to be a function of the number of EDs and Gws in the network. We also show that with our scheme, the packet loss rate is almost zero.

2 Related Work

Previous studies in the use of LoRa and real-time (RT) communications [1,3,20,21,25,26] have focused on exploring the RT capabilities of the LoRa protocol, proposing new MAC layers on top of LoRa. Other protocols [23,27] have proposed slotted communications based on the MAC layer to study application deadlines and only consider a single Gateway.

Similar to ours, in [15,24,25] the proposals are compatible with LoRaWAN introducing some kind of slotting or changes in the ACK mechanisms. In these proposals, only one single gateway is used, which is not scalable.

A new approach adds a duty cycle-aware real-time scheduling algorithm [2] to schedule links using the LLF (Least Laxity First) real-time scheduling algorithm, while minimizing the duty-cycle. However, this algorithm is not compatible with the LoRaWAN specification and it is not clear how the scheduling works with multiple gateways. In [13] the authors proposed an integer linear programming model to allocate EDs to gateways using different channels and SFs minimizing the amount of gateways. In [8], the authors proposed a fully compatible LoRaWAN scheduling algorithm for real-time messages.

The main distinctions of our proposal is that it is (a) fully compliant with the LoRaWAN specification and (b) scalable to multiple Gws. LoRaWAN compatibility is crucial to preserving the benefits of using this widely adopted and robust technology, as deviation from it would risk undoing years of testing and development of the protocol. Moreover, an implementation on top of the unchanged MAC layer can take advantage of the huge amount of devices and libraries that are already available to set up LoRaWAN networks. Scalability is essential in large-scale applications, such as precision agriculture.

3 Models and Problem Description

3.1 Assumptions

The key LoRaWAN network assumptions for this work are: (a) EDs are LoRaWAN Class A [5], fixed (not mobile) in the network, and only transmit data periodically; (b) There is no re-transmission mechanism; (c) EDs have synchronized clocks among themselves [22], so when they transmit a packet in a slot, the transmission does not overlap with another transmission in the same channel and SF; (d) EDs connect to the LoRaWAN network by ABP (Activation by Personalization, which provides fixed addresses to the EDs) [18].

For clock synchronization, each ED has an RTC clock and Gws periodically transmit a beacon with a time reference (similar to Class B devices). Consequently, our time slots contain time for beacon reception. Alternative clock synchronization techniques could be integrated into this proposal (e.g., [22]).

Our scheduling is done *offline* by a program that knows each ED's GW, SF, and temporal constraints. The schedule can be either provided manually by a network administrator or by a process that can access data in the Network Server database. Additionally, we assume once the schedule is built, it is distributed to all EDs.

3.2 Network and Communication Model

We consider a LoRaWAN network composed of a set of N Class A EDs and a set of M Gws that operate under European regulation constraints[1]. The following notation is adopted for the algorithm: $ED_{i,j}$ denotes ED i transmits with spreading factor j, $j \in \{7, 8, 9, 10, 11, 12\}$.

The EDs and the Gws communicate in the network as follows: (a) An ED does not communicate with any other ED in the network; (b) As in a typical LoRaWAN network each ED can communicate with one or more GWs; (c) Each

[1] We pick a random region to illustrate our mechanism; our description is generic and can be applied to another region.

$ED_{i,j}$ in the network can transmit in only one Spreading Factor j. (d) Time is discretized into slots and each ED uses one slot to transmit at a time.

To avoid collisions in the network, knowing that channel selection cannot be controlled, the nodes (EDs and Gws) must be under a communication model that guarantees that any two nodes not sufficiently away from each other use a different combination of channel, spreading factor, and slot to transmit a message.

3.3 Problem Description

Although LoRaWAN is a promising candidate for IoT LPWAN, its adoption may be in jeopardy given the time-sensitive needs of real-time applications. This major issue is due to the use of the Aloha-based Medium Access Control (MAC) layer, where devices transmit without any carrier sensing. In fact, LoRaWAN relies on Aloha to guarantee cost-effective communications by avoiding MAC overhead and reducing the complexity of the EDs' time synchronization within the limitations of duty-cycle and unlicensed bands. However, Aloha does not offer good reliability and leads to very poor scalability. In Aloha, transmissions have a high probability of collisions that increase packet loss, reduce throughput and increase latency due to re-transmission. Two other factors can increase collisions: high network density and the LoRaWAN adaptive SF scheme.

In a previous work, we proposed an RT-scheduling algorithm based on the Earliest Deadline First policy (EDF) that gave great results: 0% deadline miss with very low packet loss rate [8]. However, the EDF-based scheduler does not scale well since, when the number of Gws increases and the number of EDs that communicate with many Gws increases, the number of EDs that cannot be scheduled without creating collision risk also increases (see Fig. 3). For this reason, we propose a new method based on the graph-coloring technique. This not only schedules EDs properly ensuring no collisions, but also scales very well, as we can always find a proper number of colors (slotting). Like the previous proposal, this method is also built on top of LoRaWAN to avoid modifying this protocol and to be fully compatible with it.

Although RT scheduling can prevent collisions and drastically reduce packet loss, it does not completely eliminate it. In fact, it is important to consider that collisions are not the only cause of packet losses in a LoRaWAN network. Since it uses Aloha, packet loss can easily be mistakenly attributed to a collision. But packet loss can also be from radio frequency interference and weak signal (due to distances or multi-path fading).

4 LoRaWAN Graph Coloring-Based RT-Scheduling

In this work, we propose an algorithm based on graph coloring (GC) to schedule EDs with Real-Time (RT) restrictions in LoRaWAN, to avoid collision in the network. The scheduling algorithm is built on top of the LoRaWAN MAC layer, so no changes and no additional communication rules are added to the internal LoRaWAN mechanism/protocol.

From the point of view of the specification, this proposal runs in the application layer of a LoRa ED. Figure 1 illustrates this stack. Neither the gateway nor the network server need to be modified.

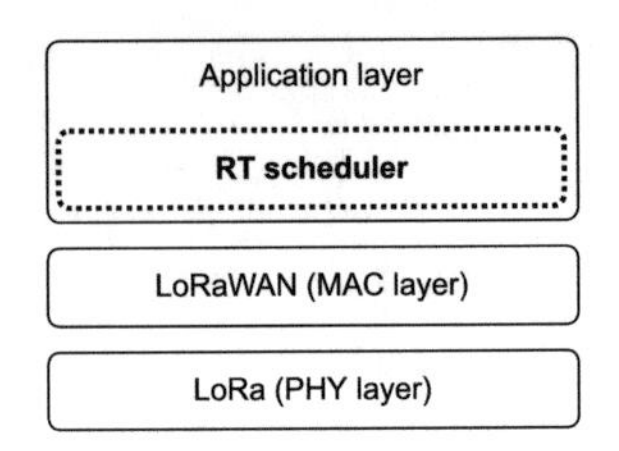

Fig. 1. LoRaWAN protocol stack of an End-Device with RT scheduling.

Our RT scheduling algorithm is based on the network communication model and assumptions described above. The RT scheduler will be run by a program to produce a schedule that will be sent to the EDs. The algorithm aims to schedule all the EDs in the network so that they can communicate without collisions and without missing their deadlines.

The algorithm performs the following steps to schedule ED transmissions:

1. **Create SF groups.** The network ($Network = \{EDs, GWs\}$) is first divided into SF groups S_j (one per spreading factor j), so that each group contains all the EDs that transmit in the same SF; $S_j = \{ED_{i,j},\ 1 \leq i \leq N\}$. Note that the SF for each node is assigned in the firmware configuration or by the NS ahead of the time; we assume that the SFs are pre-determined.
2. **Create graphs.** For each SF group S_j, we create a graph with possibly several disjoint components (nodes in a disconnected component do not interfere with nodes in a different component), $G_j = (V, E, \phi)$ where:
   - $V = \{ED_{i,j} \in G_j\}$, all EDs in G_j
   - $E = \{e_k\}$ the set of edges defined by ϕ

$$\phi(ED_{l,j}, ED_{m,j}) = \begin{cases} e_k = (ED_{l,j}, ED_{m,j}) & \text{if } ED_{l,j} \text{ and } ED_{m,j} \text{ transmit to the same GW.} \\ - & \text{otherwise} \end{cases}$$

3. **Color graphs in SF groups.** The coloring follows this rule: no two vertices (EDs) sharing an edge (transmitting to the same GW) have the same color. The graph coloring is done using Greedy coloring algorithm with max(number of ED per GW in G_j) + 1 number of colors. Figure 2 gives an illustration through a simple but representative network.
4. **Map color to slots.** After coloring, EDs that have the same color (that is, no edge, which means no collision) can share the same slot.

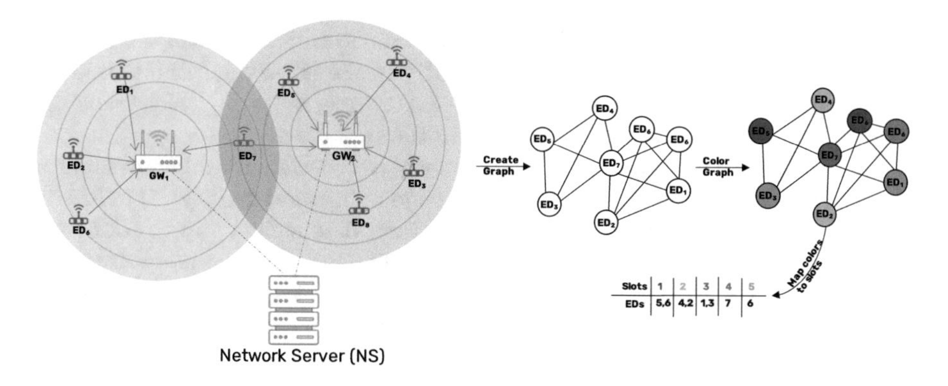

Fig. 2. Example of a network slotting steps using graph coloring.

5 Experiments

5.1 Simulation Setup

Three different experiments are conducted using a LoRaWAN module on NS-3, an open-source discrete event network simulator. The LoRaWAN module for NS-3 is a widely used tool in research studies on LoRaWAN networks [7,10,11]. To adapt the module to the present simulation, several modifications were made, such as customizing the transmission delay for each ED (needed for random LoRaWAN and calculating the delays to check deadlines), adding new callbacks to measure transmission time and packet reception, and modifying the interference helper to identify which packets and EDs were causing collisions.

All simulations are performed using different configurations of Gws and EDs. In all of them, the gateways were arranged in a grid formation [6], allowing overlap in the Gw coverage area (which allows for nodes to connect to more than one Gw). The EDs were randomly distributed according to a uniform distribution. The packet loss rate and missed deadlines are used as metrics for the experiment.

As a baseline, we create *random LoRaWAN*, which takes as input the same scenarios as our RT scheduler, but each ED decides when to send the packet using the uniform random distribution *before its deadline*. We note that we could have used as a baseline the LoRaWAN standard, but it would not be a fair comparison, given that it does not consider deadlines at all.

In the first experiment, we compared the GC-based RT-scheduling, EDF-based RT-scheduling, and the *random LoRaWAN* because we want to show the limit of the EDF-based RT-scheduler compared to the GC-based RT-scheduler. In the last two experiments we only compared with random, since the EDF scheduler has a problem scheduling all the nodes when the network scales.

In the first two experiments, all the EDs have the same periods, and in the last experiment they use different periods depending on their SF.

- **Experiment 1:** To study the scalability and reliability of our GC-based RT-scheduling algorithm, we evaluate whether the proposed solution eliminates (reduces) collisions in several LoRaWAN configurations, density, scale, etc. (see Table 1). In this experiment, the EDs of the three algorithms used the same periods for transmission. Also, we use a single SF = 12, which translates into a fixed data rate (DR0, specified in the table).

- **Experiment 2:** To evaluate the efficiency of our method for different applications, we simulate different network coverage ranges and density: (i) industrial deployment (up to 1 km), (ii) standard urban deployment (up to 5 km) and (iii) more rural applications (over 5 km). The density of the nodes increases for each coverage range, and gradually we add GW (see Table 1). As in Experiment 1, we use the SF = 12 and DR0 data rate.
- **Experiment 3:** For this experiment, our objective was to (i) obtain a more complete assessment of the GC-based scheduling algorithm that involves dividing the network into SF groups and (ii) get a better idea of whether playing with different SF configurations across distances can help achieve greater reliability and reduce packet losses that are not caused by collisions. Therefore, we evaluate the effectiveness in a heterogeneous LoRaWAN network (different SF, number of GWs, number of EDs, etc.). In this experiment, we use the LoRaWAN Adaptive Data Rate (ADR) mechanism [19] to assign the data rate (and thus periods of applications) and the SF.

To properly evaluate our collision suppression (reduction) solution, we need to distinguish packet loss caused by collisions from that caused by interference, signal attenuation due to distance, and packet drop due to weak signals. We will discuss how our results can help determine the causes of packet loss.

Table 1. Different scenario's configurations for each experiment

Exp.\Param.	GW	EDs	Data rate	Slot allocation	Network Coverage (NC)
Exp. 1	1, 2, 4, 8	100, 200, 400, 800, 1600, 3200	DR0	GC-based RT-schedule, EDF-based RT-schedule, Random	9 km
Exp. 2	1, 2, 4, 8	100, 200, 400, 800, 1600, 3200	DR0	GC-based RT-schedule, Random	Different NC varying from 1 km to 8 km
Exp. 3	1, 2, 4, 8	100, 200, 400, 800, 1600, 3200	ADR	GC-based RT-schedule, Random	9 km

6 Results and Discussion

6.1 Results

Experiment 1. In this experiment, 24 scenarios were tested. For each scenario, a number of gateways and a network density are chosen, and the packet loss and missed deadline are measured. The number of gateways varies from 1 to 8 and doubling each time. For each chosen number of gateways, we increase the network density that doubles each time from 100 to 3,200 EDs. We note that for all three methods, all EDs met their transmission deadlines.

Figure 3 shows the packet loss percentage (or rate). First, for the EDF-based RT-scheduler, even if it has a very low packet loss percentage, it does not scale well. Indeed, when the number of gateways and density increases (8 Gws and 800+ EDs), some EDs are not scheduled because scheduling them means sharing slots with EDs of the same gateway group, which will lead to collisions. Second, the GC-based RT-scheduler does not have a scalability problem, slightly improves the packet loss % of the EDF-based RT-scheduler and significantly outperforms *random LoRaWAN*. The mean packet loss % is less than 1% for the EDF and GC schedulers, and around 10% for the Random scheduler.

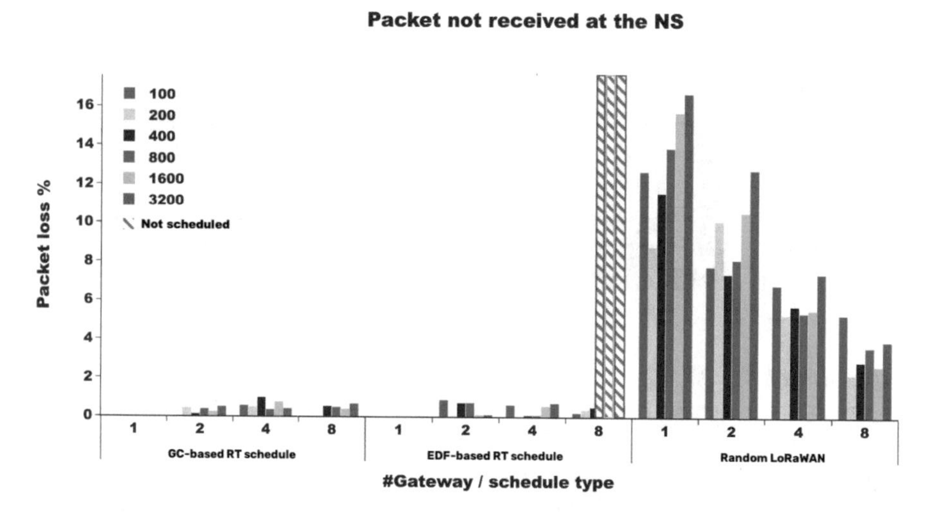

Fig. 3. Packet loss % at the Network server, comparison between the GC-based RT scheduler, the EDF-based RT-scheduler, and the *Random LoRaWAN*

Experiment 2. For this experiment, the performance of GC-based RT-scheduler is evaluated in different network coverage and density. The same scenarios as in experiment 1 are used for each network coverage. As the EDF-based RT-scheduler will not be able to schedule all EDs above a certain network density (mainly because in high densities, EDF-based RT-scheduler is too conservative), we only present our new GC scheduler (Table 2) against the *random LoRaWAN* (Table 3). The tables show that GC-based largely outperforms *random LoRaWAN*. More importantly, up to 4 km the GC-based RT-scheduler has 0% packet loss %, and above 5 km, an average of 1%, while *random LoRaWAN* varies between 10% and 15% through different network coverage.

Table 2. Percentage of packet loss for GC-based scheduling

Coverage range	1 km				2 km				3 km				4 km				5 km				6 km				7 km				8 km			
#ED \#GW	**1**	**2**	**4**	**8**	**1**	**2**	**4**	**8**	**1**	**2**	**4**	**8**	**1**	**2**	**4**	**8**	**1**	**2**	**4**	**8**	**1**	**2**	**4**	**8**	**1**	**2**	**4**	**8**	**1**	**2**	**4**	**8**
100	0	0	0	0	0	0	0	0	0	0	0	0	0	0	0	0	0	0	2	0	0	0	1	0	0	0	0	0.37	0	0	0	1.4
200	0	0	0	0	0	0	0	0	0	0	0	0	0	0	0	0	0	0	0	1	0	0	1	0	0	0	0	0.49	0	0	0.11	0.08
400	0	0	0	0	0	0	0	0	0	0	0	0	0	0	0	0	0	0	1	0	0	0	1	0	0	0	0	0	0	0	0	0
800	0	0	0	0	0	0	0	0	0	0	0	0	0	0	0	0	0	0	0	0	0	0	0	0	0	0	0	0.10	0	0	0.18	0.08
1600	0	0	0	0	0	0	0	0	0	0	0	0	0	0	0	0	0	0	0	0	0	0	0	0	0	0	0.11	0.09	0	0	0.56	0.08
3200	0	0	0	0	0	0	0	0	0	0	0	0	0	0	0	0	0	0	0	0	0	0	0	0	0	0	0	0	0	0.14	0	0.14

Table 3. Percentage of packet loss for *Random LoRaWAN*

Coverage range	1 km				2 km				3 km				4 km				5 km				6 km				7 km				8 km			
#ED \#GW	**1**	**2**	**4**	**8**	**1**	**2**	**4**	**8**	**1**	**2**	**4**	**8**	**1**	**2**	**4**	**8**	**1**	**2**	**4**	**8**	**1**	**2**	**4**	**8**	**1**	**2**	**4**	**8**	**1**	**2**	**4**	**8**
100	10	9	1	1	19	5	5	2	12	2	2	2	14	8	2	1	14	6	1	0	12	7	4	1	18	6	2	3	13	6	5	4
200	15	7	3	1	14	5	2	1	15	3	1	2	15	5	2	1	13	5	2	1	13	7	3	2	11	8	2	3	14	8	6	4
400	16	8	3	2	15	4	3	2	16	7	3	2	16	6	4	1	16	6	6	1	11	9	5	2	16	7	4	3	13	10	7	3
800	14	4	2	1	11	5	4	1	14	6	4	1	15	7	3	1	10	7	2	1	15	6	3	2	15	7	5	3	12	8	6	3
1600	16	9	2	1	17	7	3	1	17	7	3	1	16	7	4	1	18	6	4	2	18	8	4	2	18	9	5	2	16	10	6	3
3200	21	9	4	2	17	8	3	1	20	10	4	2	18	9	4	2	18	8	4	2	20	10	5	3	19	12	6	3	19	12	7	4

Experiment 3. We go farther in this experiment and test our solution in a more heterogeneous network, where nodes have different LoRa parameters combinations that may have an impact on the network reliability. The scenarios are the same as in experiment 1 but the LoRa parameters of each ED depend on the data rate allocated to the ED by the ADR mechanism. Figure 4 presents

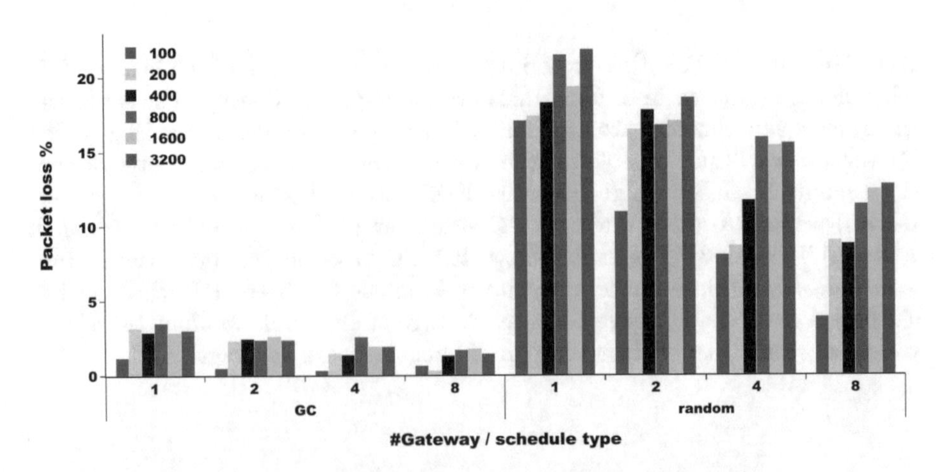

Fig. 4. Packet loss % at the Network server, comparison between the GC-base RT-scheduler and *Random LoRaWAN for scenarios with different SF*

the comparison results. They showed that GC-based still outperform *random LoRaWAN* and offers a good packet loss %, even though it is higher than in experiment 1, where the EDs use the same SF.

6.2 Discussion

Our GC-based scheduler is evaluated in extensive simulations through three different experiments where its performance is measured in different density of GWs and EDs, different network coverage, and different LoRa parameters to prove its applicability with efficiency in almost all types of LoRaWAN applications and deployments. In experiments 1 and 2, the average packet loss % after scheduling is almost zero, less than 1%. In experiment 3, the packet loss % is higher because the ADR mechanism assigns a different SF/data rate for each node, which might hinder the efficiency of the algorithm. The packet loss is due to interference (given that the GC avoids collision altogether) that occurs at the physical layer and that is not easy to eliminate by a scheduler on top of the MAC layer.

In fact, generally, in wireless communication, interference occurs in inter-network cohabitation where external signals from one network interfere with signals of the considered network. In particular, LoRaWAN has a specific network topology in which EDs broadcast their messages to gateways without specifying which should receive them. So, all gateways in the range of transmissions can receive them. This means that two LoRaWAN networks that cohabit can suffer from inter-network interference, but also in the same network, transmissions to different gateways can cause inter-LoRaWAN gateway interference. Specifically, in our scheduled network in which the EDs that are in different gateway range can send at the same time (share the same slot), all packet loss are caused by these inter-gateway interference. In fact, the experiments show that all EDs that lost packets had shared slots with EDs that are *not in the same Gw coverage but close enough* (just at the edge, around 100 m from the GW range) to cause interference. This situation occurs in LoRa because of the way LoRa treats weak signals. In fact, when two signals arrive at the same time at the gateway or overlapped in time, LoRa always tries to decode at earliest one by exploiting the benefit of the capture effect mechanism. The effect of the latter on the reception/drop of LoRaWAN packets is very detailed in [14,16].

In LoRa, the capture effect is exploited as follows: When two signals are sent in the same slot, the one received with a higher power level (at least 6 dB stronger), can still be decoded during a collision. However, interference that causes packet loss can still occur, in three specific cases: (i) when the packet with stronger signal arrives later than the receiver locking time; (ii) when the packet with stronger signal arrives after the receiver finishes receiving the weak packet signal header, and the stronger signal overlaps with the LoRaWAN header of the weaker packet; and (iii) when the stronger packet signal arrives after the receiver finishes receiving the LoRaWAN header of the weaker signal and slightly before the payload CRC of the weaker packet signal.

7 Conclusions

In this paper, we proposed a real-time graph coloring-based schedule for LoRaWANs. LoRaWAN uses the ALOHA-based MAC protocol, in which devices transmit at random times without any carrier sensing transmissions, leading to a high probability of collisions and no guarantees; thus, the standard protocol is not suitable for real-time applications that require high reliability and timeliness. In a previous work, we proposed an EDF-based RT scheduling algorithm that gave great results, 0% a missed deadline with a very low packet loss % in different scenarios, but does not scale well. For this reason, we proposed a new method based on the graph-coloring technique built on top of the MAC layer (LoRaWAN) in the application layer of the End Devices.

The performance of the real-time scheduling algorithm is assessed with the LoRaWAN module of a NS-3 simulation in different experiments for different node densities, different network coverages, and different LoRa parameters. Performance evaluation is done using two metrics: missed deadlines and packet loss rate at the destination node. In all experiments, our GC-based RT schedule and random LoRaWAN deal perfectly with sending packets before their deadlines. In addition, for our graph-coloring algorithm, the average packet loss after scheduling is almost zero, less than 1% against an average packet loss rate between 10% and 15% through the experiments for a LoRaWAN scheme that enforces sending packets before the deadline in the End Devices. The performance of the GC-based scheduler allows us to avoid collisions.

In future work, we will investigate on how to mitigate the inter-gateway interference and build a complete scheduler for LoRaWAN in order to achieve collision-free and interference zero transmissions. We will also evaluate our proposition in a real LoRaWAN deployment.

References

1. Abdelfadeel, K.Q., Zorbas, D., Cionca, V., Pesch, D.: *free*–fine-grained scheduling for reliable and energy-efficient data collection in Lorawan. IEEE IoT J. **7**(1), 669–683 (2019)
2. Afhamisis, M., Palattella, M.R.: SALSA: a scheduling algorithm for LoRa to LEO satellites. IEEE Access **10**, 11608–11615 (2022)
3. Alenezi, M., Chai, K.K., Alam, A.S., Chen, Y., Jimaa, S.: Unsupervised learning clustering and dynamic transmission scheduling for efficient dense Lorawan networks. IEEE Access **8**, 191495–191509 (2020)
4. Alliance, L.: Lorawan specification (2019)
5. LoRa Alliance. Lora and Lorawan (2023). https://lora-developers.semtech.com/documentation/tech-papers-and-guides/lora-and-lorawan/
6. Booth, L., Bruck, J., Franceschetti, M., Meester, R.: Covering algorithms, continuum percolation and the geometry of wireless networks. Ann. Appl. Probab. **13**(2), 722–741 (2003)

7. Capuzzo, M., Magrin, D., Zanella, A.: Confirmed traffic in Lorawan: pitfalls and countermeasures. In: 2018 17th Annual Mediterranean Ad Hoc Networking Workshop (Med-Hoc-Net), pp. 1–7 (2018)
8. Finochietto, J.M., Dieng, O., Mosse, D., Santos, R.M.: Adding empirical real-time guarantees to Lorawan. In: 31st ACM International Conference on Real-Time Networks and Systems, RTNS 2023, pp. 99–107 (2023)
9. Ismail, D., Rahman, M., Saifullah, A.: Low-power wide-area networks: opportunities, challenges, and directions. In: 19th International Conference on Distributed Computing and Networking, pp. 8:1–8:6. ACM (2018)
10. Magrin, D., Capuzzo, M., Zanella, A.: A thorough study of Lorawan performance under different parameter settings. IEEE IoT J. **7**(1), 116–127 (2020)
11. Magrin, D., Centenaro, M., Vangelista, L.: Performance evaluation of LoRa networks in a smart city scenario. In: 2017 IEEE International Conference on Communications (ICC), pp. 1–7 (2017)
12. Mekki, K., Bajic, E., Chaxel, F., Meyer, F.: A comparative study of LPWAN technologies for large-scale IoT deployment. ICT Exp. **5**, 1–7 (2018)
13. Micheletto, M., Zabala, P., Ochoa, S.F., Meseguer, R., Santos, R.: Determining real-time communication feasibility in IoT systems supported by LoRaWAN. Sensors **23**(9), 4281 (2023)
14. Pham, C., Ehsan, M.: Dense deployment of LoRa networks: expectations and limits of channel activity detection and capture effect for radio channel access. Sensors **21**(3), 825 (2021)
15. Polonelli, T., Brunelli, D., Benini, L.: Slotted ALOHA overlay on LoRaWAN - a distributed synchronization approach. In: 2018 IEEE 16th International Conference on Embedded and Ubiquitous Computing (EUC), pp. 129–132 (2018)
16. Rahmadhani, A., Kuipers, F.: When LoRaWAN frames collide. In: Proceedings of the 12th International Workshop on Wireless Network Testbeds, Experimental Evaluation & Characterization, pp. 89–97 (2018)
17. Santos, R., et al.: Real-time communication support for underwater acoustic sensor networks. Sensors (Switzerland) **17**(7), 1629 (2017)
18. Selimović, N.: The things stack, November 2021. https://www.thethingsindustries.com/docs/devices/abp-vs-otaa/
19. Semtech: Understanding ADR (2023). https://lora-developers.semtech.com/documentation/tech-papers-and-guides/understanding-adr/
20. Shayo, E.I., Abdalla, A.T., Mwambela, A.J.: Dynamic multi-frame multi-spreading factor scheduling algorithm for LoRaWAN. J. Electr. Syst. Inf. Technol. **10**(1), 11 (2023). https://doi.org/10.1186/s43067-023-00077-2
21. Sisinni, E., et al.: Enhanced flexible LoRaWAN node for industrial IoT. In: 14th IEEE International Workshop on Factory Communication Systems (WFCS) (2018)
22. Wadatkar, P., Zennaro, M., Manzoni, P.: On time synchronization of LoRaWAN based IoT devices for enhanced event correlation. In: 6th EAI Internaitonal Conference on Smart Objects and Technologies for Social Good, GoodTechs 2020 (2020)
23. Xu, Z., Luo, J., Yin, Z., He, T., Dong, F.: S-MAC: achieving high scalability via adaptive scheduling in LPWAN. In: IEEE Conference on Computer Communications, IEEE INFOCOM 2020, pp. 506–515 (2020)
24. Yapar, G., Tugcu, T., Ermis, O.: Time-slotted ALOHA-based LoRaWAN scheduling with aggregated acknowledgement approach. In: 2019 25th Conference of Open Innovations Association (FRUCT), pp. 383–390. IEEE (2019)
25. Zorbas, D., Abdelfadeel, K., Kotzanikolaou, P., Pesch, D.: TS-LoRa: time-slotted LoRaWAN for the industrial internet of things. Comput. Commun. **153**, 1–10 (2020)

26. Zorbas, D., Abdelfadeel, K.Q., Cionca, V., Pesch, D., O'Flynn, B.: Offline scheduling algorithms for time-slotted LoRa-based bulk data transmission. In: IEEE 5th World Forum on Internet of Things (WF-IoT) (2019)
27. Zorbas, D., O'Flynn, B.: Collision-free sensor data collection using LoRaWAN and drones. In: 2018 Global Information Infrastructure and Networking Symposium (GIIS), pp. 1–5 (2018)

CertifIoT: An IoT and DLT-Based Solution for Enhancing Trust and Transparency in Data Certification

Francisco Moya, Francisco J. Quesada(✉), Luis Martínez, and Fco Javier Estrella

Universidad de Jaén, Campus Las Lagunillas, 23071 Jaén, Jaén, Spain
{fpmoya,fqreal,martin,estrella}@ujaen.es

Abstract. Data accuracy, reliability, and integrity are of utmost importance in multiple scenarios, such as medical test, performance testing, etc.; data certification plays a critical role in providing trust, as it ensures that the data is guaranteed by trusted entities. However, traditional private and centralised approaches make it challenging to guarantee the veracity of the data or establish liability in cases of deliberate manipulation or human error.

This contribution introduces *CertifIoT*, a Distributed Ledger Technology (DLT)-based solution for certifying data gathered from data streams produced by Internet of Things (IoT) devices. This approach adds transparency to the process and provides a mechanism for clarifying data provenance and recording methods. The foundation of our solution is the Phonendo Framework [1], which has been extended with additional features required for data certification. Furthermore, we propose an adaptable and extensible data model, to make certified data interoperable, and a protocol for trust endorsement, which aims at increasing the trust of certified data.

Keywords: Certification · Trust Endorsement · IoT · DLT · Data Model

1 Introduction

In today's data-driven systems era, data accuracy, reliability, and integrity are crucial, especially in critical scenarios such as performance testing of athletes, information exchange between public/private health systems for diagnostic purposes, or validating the purchase of health insurance policies. Inaccurate or unreliable data can have undesirable consequences, including misdiagnoses, inappropriate treatment decisions, compromised athletic fairness, and unfair competition [2]. Therefore, ensuring trust in the data becomes imperative to maintain the integrity and effectiveness of these scenarios.

In this context, data certification plays a pivotal role in establishing trust and credibility in data-driven systems [3]. By certifying the data, stakeholders can

J. Bravo and G. Urzáiz (Eds.): UCAmI 2023, LNNS 841, pp. 127–138, 2023.
https://doi.org/10.1007/978-3-031-48590-9_12

have confidence that it is guaranteed by given entities and that meets specific standards of accuracy and reliability.

Traditional private and centralised approaches to data management face significant limitations when it comes to ensuring data accuracy, reliability, and integrity. These approaches often lack transparency, making it challenging to verify the authenticity of the data or trace its origin and capture method [4]. Moreover, they frequently lack robust mechanisms to establish liability in cases of deliberate manipulation or human error, leaving stakeholders vulnerable to the consequences of unreliable data. These limitations highlight the need for innovative decentralised approaches that can effectively address these challenges [5].

The emergence of DLTs has fostered the development of proposals that promote the management of certificates in a decentralised manner, given that their properties encourage process transparency and access to information. Numerous proposals can be found, for instance, in the academic sphere for issuing diploma certificates [6] or to generate and verify health certificates in a hospital [7].

In addition, to make certificates valuable, it is necessary to trust the entity that issues them. This entails that stakeholders need to know the identity of the issuer and decide whether the issuer is trustable or not. DLTs have also been used to manage identities in a decentralised manner [8]. Kurbatov et al. proposed a decentralised system to create a decentralised public key infrastructure with flexible management for user identifiers [9], while Fan et al. presented "Diam-iot" a decentralised framework for identity and access management in IoT environments [10].

Since the expansion of the Internet, researchers have been working to address the issues associated with identity on the Internet (e.g. identity theft or password breaches). Traditionally, user data has been stored on organisations' servers, leading to a loss of control over personal information by users and potential risks if the security of those servers is compromised [11]. For this reason, in recent years, efforts have been focused on regaining control of one's digital identity, in what is known as Self-Sovereign Identity (SSI) [12]. Indeed, the publication of various W3C[1] specifications for defining their key elements, such as the Verifiable Credentials (VCs) [13] or the Decentralised Identifiers (DIDs) [14], has fostered the proposal of new SSI solutions in different scenarios [15,16]. However, elements such as VCs are designed for data that are not continuously changing (e.g. a passport). Consequently, it is necessary to explore how can we address data certification in IoT ecosystems that generate data streams.

This contribution proposes a novel DLT-based solution for certifying data from IoT environments. The solution extends the Phonendo framework [1] with essential features for data certification. Additionally, the paper introduces an adaptable and extensible data model, along with a protocol to foster trust between entities who participate in the certification process.

The remaining of the contribution is structured as follows. In Sect. 2 the proposed certification process is detailed, considering its elements and partici-

[1] https://www.w3.org.

pants. Section 3 describes the considerations to take into account when applying data certification into IoT. Section 4 introduces CertifIoT, highlighting its architecture, data flow and implementation details. An illustrative example of IoT data certification is presented in Sect. 5. Finally, Sect. 6 summarises the total contribution and identifies future research opportunities.

2 Background on the Certification Process

This section reviews the basics of the certification process, focussing on their main elements and participants.

Definition 1. *A credential or certificate can be defined as a tamper-resistant collection of information that is asserted as true by an authoritative entity [12].*

The trustworthiness of this authority plays a pivotal role, as it determines the level of trust placed in the information encapsulated within the credential. An example is a university-issued diploma, which serves as tangible evidence of an individual's educational attainment. Each certificate comprises a series of claims related to the subject that include attributes, relationships, or entitlements. The verifiability of these claims is essential, necessitating the verifier's ability to ascertain the identity of the issuer, detect any signs of tampering, and ensure that the certificate remains valid by verifying its non-expiration or non-revocation status. Therefore, each claim should be proved by its associated *proof/s*, which demonstrate the claim is true.

Considering the participants within a certification process we can distinguish between [12]:

- *Issuer*, who emits the given certificate.
- *Holder*, who requests certificates from issuers, securely retains them, and provides proof of claims associated with one or more certificates when requested by verifiers.
- *Verifier*, who might be an entity (individual, organisation, or an automated algorithm) that seeks assurance of trust regarding the subjects of certificates.

The relationship between issuer, holder, and verifier is known as the *trust triangle* (see Fig. 1).

Thus, a certificate is only accepted by a verifier if the holder presents its proof/s, and it is issued by a trusted issuer whom the verifier trusts. Otherwise, if the verifier does not have trust in the issuer or is unable to validate the proof/s, the certificate will not be accepted. For instance, consider a scenario where a student has recently completed a bachelor's degree at University A and applies to enrol in a master's program at University B. If University B does not trust University A, the bachelor diploma will not be considered valid, resulting in the student being unable to enrol in the master's program at University B.

In order to allow verification of digital identity the W3C proposed the concept of DIDs [14], which serve as a novel form of identifier that facilitates verifiable

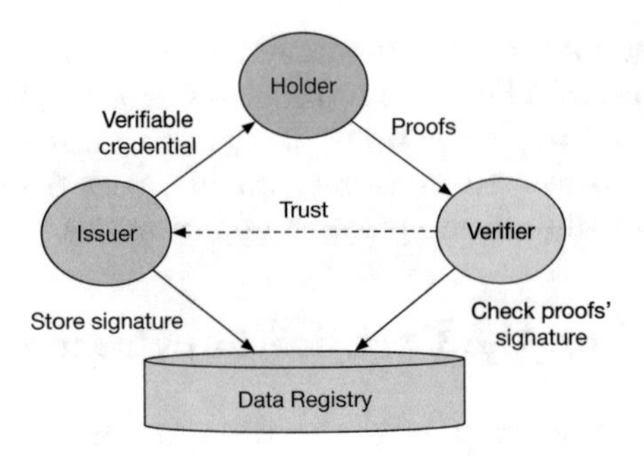

Fig. 1. Trust triangle (adapted from [12]).

and decentralised digital identity. DIDs are stored in a data registry, accessible by verifiers, and serve as unique identifiers that represent individuals, organisations, devices, or abstract entities. They are uniform resource identifiers (URIs) that connect a DID subject to a DID document, enabling trustworthy interactions. Thus, DIDs possess the following characteristics: (i) *permanence*, ensuring they remain unchanged over time; (ii) *resolvability*, enabling the retrieval of associated data; (iii) *cryptographic verifiability*, allowing identity holders to prove their control over the private key associated to the identifier using cryptographic methods; and (iv) *decentralisation*, mitigating the risk of single point of failure.

The emergence of DLTs has provided an excellent opportunity to utilise them as DIDs registries, due to their alignment with the characteristics of DIDs.

Another key aspect is the understanding of the content encapsulated within a certificate or credential. Thus, the VC Data Model 1.1 specification [13] defines data models for VCs, verifiable presentations and how to represent data models with JSON Linked Data (JSON-LD) [17] and JSON Web Tokens (JWT) [18].

Despite the well-known elements of SSI, they are designed to deal with a certain amount of data that are not continuously produced or changing (e.g., academic diploma, access card, etc.). In order to allow data certification in IoT environments, we propose an approach capable of certifying data from data streams.

3 Applying Data Certification into IoT

The application of data certification to IoT data streams may enhance the overall security, reliability, and transparency of IoT data, promoting the responsible and trustworthy use of IoT technologies. However, several issues must be addressed, for example:

- How can certified data be made interoperable, ensuring it is understandable and equally interpreted by different entities?
- What approaches can be employed to establish a network of trusted entities?
- How can we efficiently retrieve data certified by a specific entity?

Below, it is provided further detail about the approaches carried out in our research to address these questions.

3.1 Data Models Creation for Interoperability

Any VC or certificate contains data about the claims, the issuer, the subject for whom the VC has been issued, the validity period or expiration date of the VC, and whether it has been revoked or not.

Regarding interoperability and data sharing, it is necessary to define data models that allow data interpretation. Thus, we propose using data models with the aim that any stakeholder can verify the claims/proofs, considering the data formatted in the given data model. This allows us to model data, making easier its capture, processing and verification, while fostering interoperability.

Additionally, by defining a data model schema and referencing it within the captured data, we minimise the DLT data load because it is not necessary to include in every new record the name of the attributes represented, as the data can be interpreted using the linked data model schema. This reduces the size of the records and data flows, which is key in IoT.

In our proposal, data models can be stored in both a DLT or a web server. In both cases, immutability is guaranteed, which is essential to keep data consistency. However, the latter allows more flexibility (e.g. restricted access), but it is necessary to provide a prove that guarantees the data model has not been changed. For this reason, we store in the DLT the hash of the data model stored in the web server.

Therefore, in order to enhance data interoperability, we propose a data model following the JSON schema format with the following fields:

- *name*: name of the schema.
- *namespace*: URL to the document of the specification used.
- *version*: schema version.
- *description*: brief description the of the schema representation.
- *extends*: URL of the schema that is extended (optional).
- *fields*: List of fields defined in the schema. Each field is an object with *name*, *type* and *description*.

3.2 Establishment of a Trust Chain

Apart from interoperability issues, another critical concern in certification is the trusted network, because if verifiers do not trust issuers, the issued certificates will not be considered valid. For this reason, it is proposed allowing participants endorsement in order to create robust trusted networks.

To do so, verifiers' signatures need to be published in the DLT including their public keys and profiles. This assures authenticity and provides transparency to the process (see Listing 1).

Listing 1. Example of verifier signature.

```
1  {
2       "organisation": "Name",
3       "publicKey": "VERIFIER_PUBLIC_KEY",
4       "profile": <dlt_name:tx|web:url:hash>,
5  }
```

Similarly to data models, the identities of verifiers can be store in a DLT or a web server. The link to the resource should be included in the "*profile*" field.

In order to improve the trust of certificates, verifiers can endorse each other. This increases data reliability and reduces the risk of fraud or data manipulation. To conduct an endorsement it is necessary to carry out a transaction to the DLT including: (i) the public key of the verifier who endorses, (iii) the public key of the endorsed verifier, and (iii) other metadata such as the endorsement validity (see Listing 2)

Listing 2. Example of verifier endorsement.

```
1  {
2       "endorserKey": "ENDORSER_PUBLIC_KEY",
3       "endorsedKey": "ENDORSED_PUBLIC_KEY",
4       "metadata": {
5           "validityPeriod": "EXPIRATION_TIMESTAMP"
6       }
7  }
```

As a result, it is created a *chain of trust* that can be verified by any stakeholder.

3.3 Data Retrieval Process

Once data models and verifier identities are published, the data is stored according to the corresponding data model schema, and verifiers had endorsed each other, the next challenge lies in retrieving the stored data. To address this challenge, we propose a review and validation process for the received data, based on the trust chain established by the endorsement process.

1. *Data and schema retrieval.* When new input data are received, the associated schema is first retrieved using the reference of the schema included in the data. The verifier uses this reference to search for the schema at the specified location (DLT or web server).
2. *Data composition.* Once the schema is retrieved, the input data are composed according to it. This involves recognising the data structure and ensuring its compliance with the respective schema.
3. *Trust chain retrieval.* To retrieve the verifier's trust chain, it is necessary to search for endorsements in the network that points to the verifier. Thus, the verifier can follow this trust chain until reaching a trusted entity, indicating the reliability of the data.
4. *Data verification.* Finally, the received data are validated according to the trust chain and the retrieved schema. This process ensures the accuracy and authenticity of the data.

The advantage of this validation process is that it is transparent and auditable. Each action of this process can be recorded on the DLT, providing an immutable audit trail. It strengthens the reliability of the system and allows tracking and attributing any attempted fraud or data manipulation to its source. This form of verification offers a robust and flexible solution to ensure the accuracy and authenticity of data in a wide range of use cases.

4 Overview of CertifIoT

This section presents *CertifIoT* that is a system for certifying data from IoT streams. Here, detailed information is provided about its architecture, data flow, and implementation.

4.1 System Architecture

CertifIoT extends Phonendo's architecture [1] with the required modules to enable data certification. Figure 2 illustrates the different components and their interactions[2].

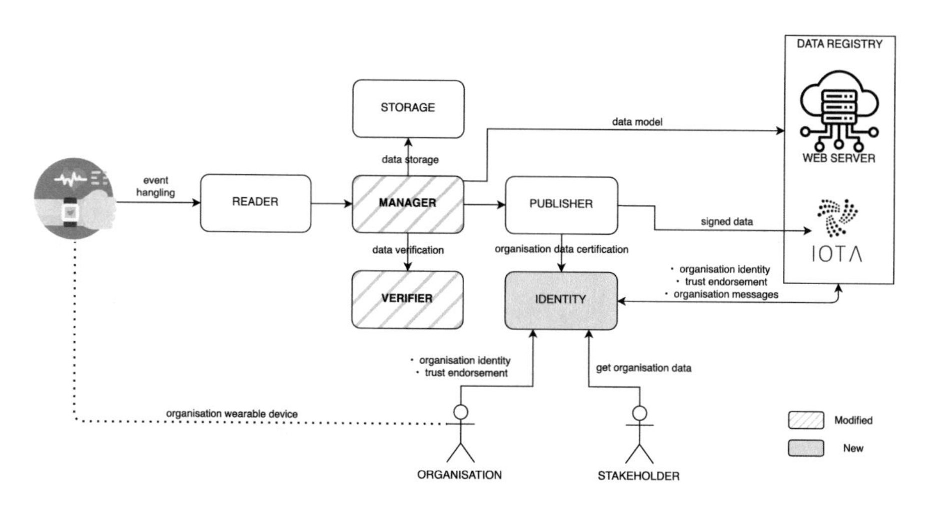

Fig. 2. CertifIoT Architecture.

The main extension is the development of the module called "*Identity*", which is responsible for the following tasks:

[2] It is necessary to highlight that the term "Verifier" in Phonendo's architecture refers to a service that verifies the received data before signing it. Therefore, it is not the same as a *verifier* in the certification process. For the sake of clarity, every reference to *verifier* in the text refers to the latter, a relying entity.

- *Identity management.* The process involves using the organisation's public key to build a transaction with the CertifIoT identity structure and publishing it on the data registry with an index that allows retrieval of the published data.
- *Trust endorsement.* Allowing entities involved in the process to endorse each other. This provides confidence and robustness to the data published by an endorsed organisation.
- *Data retrieval.* Retrieving all data published on IOTA by a given organisation. This includes the organisation's identity, the entities who endorsed the given organisation, and all data published by all of the organisation's devices.

The integration of the *Identity* module within Phonendo also entailed extending existing Phonendo modules. The *Phonendo Manager* and *Phonendo Verifier* components have been enriched with new functionality. The former can now obtain data models from the data registry, while the latter allows organisations to sign their data.

4.2 Data Flow

Figure 3 depicts the normal sequence diagram of the proposed system[3].

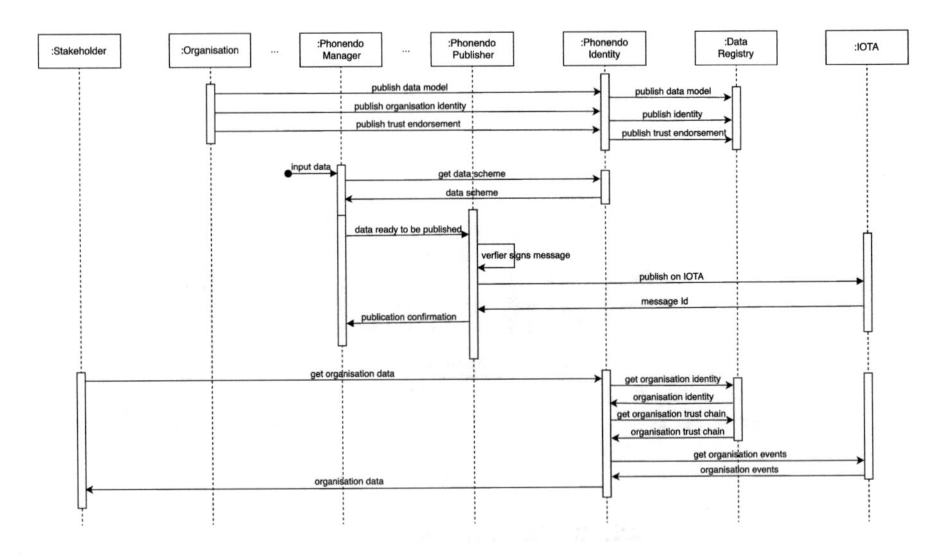

Fig. 3. Sequence diagram.

First of all, the organisation needs to make its identity public. To do so, it uses *Phonendo Identity*. Similarly, it can also endorse other entities.

Secondly, when *Phonendo Manager* receives new input data, it retrieves the corresponding data model from the data registry and formats the input data

[3] Notice that some parts of Phonendo's data flow have been omitted to focus on the proposal.

according to the obtained data model. The formatted data is then published on IOTA after *Phonendo Identity* has signed it with the organisation's signature.

Finally, a stakeholder can consult specific certified data published on IOTA by interacting with *Phonendo Identity*.

4.3 Implementation Details

The code developed for this project is publicly available[4] and focused on 3 main points: (i) identity management, (ii) trust endorsement (iii) and data publication/retrieval. Regarding identity management (i), *Identity*, exposes several HTTP endpoints to get/publish the organisation's identity from/to IOTA. The publication of the identity on IOTA follows a common index definition to be dynamically retrieved in real-time with the pattern `PHONENDO:IDENTITY:ORGANISATION` from IOTA.

Concerning trust endorsement (ii), an HTTP method has been developed to allow trust endorsement between two different entities, resulting in more confidence in the data provided by the organisation. Specifically, a common index definition is used with the pattern `PHONENDO:IDENTITY:TRUST:ORGANISATION`, which allows us the get all the endorsements received by an organisation.

Finally, the implementation about the publication process (iii) focused on 2 different parts: (a) Phonendo data model, and (b) data publication

For the former, once the microservice starts, it retrieves the data model schema from the data registry with the format `PHONENDO:DATAMODEL:SCHEMA_ID`, checks its *hash* and stores it in the instance. Later, the *Manager* uses the schema to transform every data event to the same structure. Regarding the latter, once the event is ready to be published, it is sent to the *Publisher* and *Identity* components. At this stage, the *Identity* component uses the public key provided by the organisation to add its signature and publish it on IOTA with the following index pattern: `PHONENDO:ORGANISATION`. This process allows obtaining all the data published by an organisation and checking the authenticity and ownership of the data.

5 Illustrative Example

Below it is presented an illustrative example to show CertifIoT performance from edge to edge. To do so, we have focused on recording heart rate data to certify stress tests. This is applicable in scenarios where a medical assessment is required to certify a suitable health condition. For instance, it can be used for specific job positions or when applying for life or health insurance.

For data collection, we have used the *Pine Time*[5] smartwatch to capture users' pulse. This wearable device is compatible with Phonendo, so no additional configurations were needed.

[4] https://github.com/sinbad2-ujaen/certifiot.

[5] https://www.pine64.org/pinetime/.

In order to make the captured data interoperable and interpretable we have defined the "*Heart rate*" data model and stored it on IOTA (see Listing 3).

Listing 3. *"Heart rate"* data model.

```
{
    "name": "heart_rate",
    "namespace": "https://phonendo.org/schema/v1.0",
    "version": "1.0",
    "description": "Schema for heart rate data",
    "extends": null,
    "fields": [
    {
        "name": "heart_rate",
        "type": "integer",
        "description": "The heart rate in beats per minute"
    },
    {
        "name": "timestamp",
        "type": "datatime",
        "description": "The time at which the heart rate was
            measured"
    },
    {
        "name": "mac",
        "type": "string",
        "description": "The MAC device where the heart rate was
            measured"
    } ]
}
```

Due to the data model is stored on the IOTA Tangle, to retrieve it, each new transaction must include the *id* of the IOTA transaction of the schema.

The process starts when the wearable device starts receiving data. Then it sends the captured data to the *Manager* in a basic/raw format.

The *Manager* has the responsibility of interpreting and formatting the captured data according to the corresponding data model. Thus, the device sends the data as shown in Listing 4, where *hr* is the heart rate, *time* is when the data was captured and *mac* is the MAC address of the device.

Listing 4. Example of the data sent by the wearable device to Phonendo.

```
{
    "type": "HEART_RATE",
    "value": "80",
    "mac": "B5-D0-1A-98-82-02"
}
```

The *Manager* understands the data and contextualised them adding the corresponding data model to heart rate (see Listing 5).

Listing 5. Example of contextualised heart rate data according to the data model.

```
{
    "schemaID": "iota:f4fdf65d7635108cafa062c458987b625e763514420220
        1f43a39681d13596da",
    "data": {
      "heart_rate": "80",
      "timestamp": 1690215457253,
      "mac": "B5-D0-1A-98-82-02"
    }
}
```

Due to our data model is stored on IOTA, the reference to the schema "*schemaID*" includes the *iota* prefix and the *id* of the IOTA transaction[6].

The capture, formatting and publishing process continues until no new input data is received.

When a stakeholder wants to verify the data certified by a particular organisation, it is necessary to specify the organisation's *id* to *Identity*. After that, the certified data and the data models associated to these data are retrieved to the verifier. As data validation is based on the trust chain, the organisation's endorsements are also returned to the verifier. Subsequently, if the verifier trusts any endorser of the organisation's endorsements, the retrieved data is validated.

6 Conclusions

In this contribution, we have introduced CertifIoT, a novel DLT-based solution for certifying data from IoT ecosystems. By leveraging CertifIoT, we have demonstrated the potential to achieve transparency in the data certification process, enabling clear data provenance and recording methods. Our solution enhances the value of the obtained results by making them exportable and comparable, while also providing a means to detect potential data manipulations and take appropriate actions.

To develop CertifIoT, we have extended the Phonendo Framework with the *Identity* module which, implements the additional features required for data certification. We have also proposed an adaptable and extensible data model, along with a protocol for trust endorsement.

Future research will focus on supporting JSON-LD for data models and enriching the endorsement process to allow certificate revocation. Additionally, we plan to explore how to support different data models within the same IoT environment and how to notify verifiers of data model updates. Finally, we also consider exploring the application of this research in health scenarios to monitor patients during their treatments.

References

1. Moya, F., Quesada, F.J., Martínez, L., Estrella, F.J.: Phonendo: a platform for publishing wearable data on distributed ledger technologies. Wirel. Netw. (2023). https://doi.org/10.1007/s11276-023-03458-7
2. Botha, M., Botha, A., Herselman, M.: Data quality challenges: a content analysis in the e-health domain. In: 2014 4th World Congress on Information and Communication Technologies, WICT 2014, pp. 107–112. IEEE (2014)
3. Feghhi, J., Feghhi, J., Williams, P.: Digital Certificates. Addison-Wesley, Pearson Education (1999)
4. Zhang, J., Zhong, S., Wang, T., Chao, H., Wang, J.: Blockchain-based systems and applications: a survey. J. Internet Technol. **21**(1), 1–14 (2020)

[6] For data models stored in a web server, "*schemaID*" includes the *web* prefix, the schema URL and a *hash* of the schema, in order to guarantee the integrity.

5. Rahman, M.S., Alabdulatif, A., Khalil, I.: Privacy aware internet of medical things data certification framework on healthcare blockchain of 5G edge. Comput. Commun. **192**, 373–381 (2022)
6. Guustaaf, E., Rahardja, U., Aini, Q., Maharani, H.W., Santoso, N.A.: Blockchain-based education project. APTISI Trans. Manage. (ATM) **5**(1), 46–61 (2021)
7. Namasudra, S., Sharma, P., Gonzalez Crespo, R., Shanmuganathan, V.: Blockchain-based medical certificate generation and verification for IoT-based healthcare systems. IEEE Consum. Electron. Mag. **12**(2), 83–93 (2022)
8. Liu, Y., He, D., Obaidat, M.S., Kumar, N., Khan, M.K., Choo, K.R.: Blockchain-based identity management systems: a review. J. Netw. Comput. Appl. **166**, 102731 (2020)
9. Kurbatov, O., Shapoval, O., Poluyanenko, N., Kuznetsova, T., Kravchenko, P.: decentralized identification and certification system. In: 2019 IEEE International Scientific-Practical Conference Problems of Infocommunications, Science and Technology (PIC S&T), pp. 507–510. IEEE (2019)
10. Fan, X., Chai, Q., Xu, L., Guo, D.: DIAM-IoT: a decentralized identity and access management framework for internet of things. In: Proceedings of the 2nd ACM International Symposium on Blockchain and Secure Critical Infrastructure, pp. 186–191 (2020)
11. Tobin, A., Reed, D.: The inevitable rise of self-sovereign identity. The Sovrin Foundation, 29 September 2016
12. Preukschat, A., Reed, D.: Self-sovereign Identity. Manning Publications, Shelter Island (2021)
13. Noble, G., Longley, D., Burnett, D., Den Hartog, K., Sporny, M., Zundel, B.: Verifiable credentials data model v1.1. W3C recommendation, W3C, March 2022. https://www.w3.org/TR/2022/REC-vc-data-model-20220303/
14. Guy, A., Reed, D., Sporny, M., Sabadello, M.: Decentralized Identifiers (DIDs) v1.0. W3C recommendation, W3C, July 2022. https://www.w3.org/TR/2022/REC-did-core-20220719/
15. Soltani, R., Nguyen, U.T., An, A.: A survey of self-sovereign identity ecosystem. Secur. Commum. Netw. **2021**, 1–26 (2021)
16. Ahmed, M.R., Islam, A.M., Shatabda, S., Islam, S.: Blockchain-based identity management system and self-sovereign identity ecosystem: a comprehensive survey. IEEE Access **10**, 113436–113481 (2022)
17. Kellogg, G., Lanthaler, M., Sporny, M.: JSON-LD 1.0. W3C recommendation, W3C, November 2020. https://www.w3.org/TR/2020/SPSD-json-ld-20201103/
18. Jones, M.B., Bradley, J., Sakimura, N.: JSON Web Token (JWT). RFC Editor (2015). https://doi.org/10.17487/RFC7519

Unifying Wearable Data: A Novel Architecture Integrating Fitbit Wristbands and Smartphones for Enhanced Data Availability and Linguistic Summaries

David Díaz-Jiménez[1(✉)], Javier Medina-Quero[2], and Macarena Espinilla-Estévez[1]

[1] Department of Computer Science, University of Jaén, Jaén, Spain
ddjimene@ujaen.es

[2] Department of Computer Engineering, Automation and Robotics, University of Granada, Granada, Spain

Abstract. The management of wearable device data faces significant challenges due to the limited availability of suitable Application Programming Interfaces (APIs). In response to this issue, we present a pioneering architecture that seamlessly integrates data from commercially available Fitbit wristbands' sensors and smartphones, resulting in improved data accessibility and advanced linguistic summaries. Our novel approach utilises cutting-edge sensors to efficiently capture and transmit user movement and heart rate data wirelessly to smartphones. A key element of our architecture involves facilitating communication with a central platform via a robust REST API. This enables us to incorporate fuzzy linguistic protoforms, empowering sophisticated data analysis techniques to be employed. Furthermore, we have developed specific applications tailored for both mobile devices and smartwatches, enabling seamless data collection and visualizations. To demonstrate the efficacy and versatility of our proposed architecture, we conducted a comprehensive case study encompassing multiple scenarios. The results of this study affirm the substantial benefits of our approach, showcasing its potential to revolutionise wearable data management and analysis. By providing a scalable and adaptive solution to the current limitations in wearable data management, our work lays the groundwork for further advancements in this field, promising to foster new research and applications in diverse domains.

Keywords: Wearable devices · Data integration · Fitbit wristbands · Linguistic summaries · Mobile data visualization

This work has been partially supported by grant PID2021-127275OB-I00, funded by MCIN/AEI/10.13039/501100011033, and by the 'ERDF - A way of making Europe'.

J. Bravo and G. Urzáiz (Eds.): UCAmI 2023, LNNS 841, pp. 139–150, 2023.
https://doi.org/10.1007/978-3-031-48590-9_13

1 Introduction

In the last years, technological advances have been revolutionizing the world of wearable devices, providing a wide variety of options for collecting data on users' health and activity [6]. As a result of these innovations, it is becoming increasingly common to find devices that constantly monitor our vital metrics, such as heart rate, sleep quality and physical activity levels [7].

The rise of artificial intelligence (AI) has played a key role in the development of new techniques to extract valuable knowledge from the data gathered by wearable devices [11,17]. Such technologies enable complex patterns in data to be analyzed and understood [14,15], opening up a world of possibilities in terms of disease detection, early diagnosis and personalized recommendations to improve health and well-being [13,18].

Another relevant aspect in the area of wearable devices is the use of fuzzy logic, a field of artificial intelligence that allows handling uncertainty and imprecision in the data [1,2,4]. This approach is particularly valuable in areas such as healthcare where it has been applied for the early detection of preeclampsia [5], to assess the level of rest [9] or to summarise heart rate streams of patients with ischemic heart disease [16], where the data can vary and are not always absolute.

Regarding wearable devices and existing solutions, some device manufacturers provide APIs to obtain health and activity data, although it is important to keep in mind that many times this data has undergone some kind of processing before being presented to the user. This can involve the transformation of the original values or the application of filtering algorithms, which prevents perform own data processing for other purposes, such as the detection of specific activities or the implementation of custom algorithms. In this sense, wearable devices that provide access to raw data are crucial to enable researchers, developers and professionals in healthcare and other fields to perform in-depth, personalized analysis. The availability of real, unmodified values provides the opportunity to harness the full potential of techniques such as artificial intelligence and fuzzy logic.

In order to address these limitations, the development of a new, advanced and scalable architecture is proposed in this contribution. This architecture will focus on the efficient collection of raw data, allowing users to access all information and providing the opportunity for customised, in-depth analysis. A highlight of this architecture will be its ability to generate linguistic summaries, which will provide a clearer and more accessible comprehension of the data streams.

To evaluate the efficacy and potential of the proposed architecture, a comprehensive case study has been meticulously conducted, comprising three distinct scenarios.

This contribution is structured as follows: Sect. 2 presents the initial concepts related with linguistic protoforms. Section 3 presents the advanced architecture for wearable device data management. Section 4 presents the proposed linguistic protoforms for evaluating data streams. Section 5 presents a case study to illustrate the capabilities of the proposed architecture. Finally, Sect. 6 presents the conclusions.

2 Background

This section reviews the fundamentals of linguistic protoforms, which are used in the approach proposed in this contribution, as well as introduce the chosen wearable device for data integration.

2.1 Linguistic Protoforms Background

In the context of fuzzy logic, linguistic variables play a fundamental role by enabling the representation and description of information in a more flexible way and closer to natural language. These variables are built from a base variable, which takes real numerical values in a given range [20].

To approximate these real values to linguistic terms, fuzzy sets are used, where each linguistic term is bound to a membership function. This membership function establishes the degree of membership of a given value to the corresponding linguistic term. The membership degree is expressed in an interval between 0 and 1, indicating the extent to which the given value conforms to the linguistic term in question [12].

Fuzzy protoforms, on the other hand, originated as a model proposed by Zadeh in the field of fuzzy knowledge [19]. These are based on fuzzy sets whose membership degrees are determined by fuzzy membership functions. These protoforms make it possible to describe and represent processed information using natural language, which facilitates the interpretation and understanding of the results obtained from fuzzy data [8].

2.2 Wearable Device

The chosen wearable device to enrich the proposed architecture is the Fitbit Versa 3, although other models may also be considered. The Fitbit Versa 3 is an advanced wearable equipped with a diverse array of sensors, enabling comprehensive data tracking. These sensors include an optical heart rate monitor that continuously measures heart rate, an accelerometer + gyroscope combo for monitoring movement, a barometer for assessing atmospheric pressure changes, a built-in GPS for precise outdoor activity tracking, and an SpO2 sensor for monitoring blood oxygen saturation levels. The choice of this device was based on its popularity, accuracy and ease of use.

3 New Proposed Architecture

This section presents the proposed architecture for managing data from wearable devices, Fig. 1. The different layers that compose the architecture will be described.

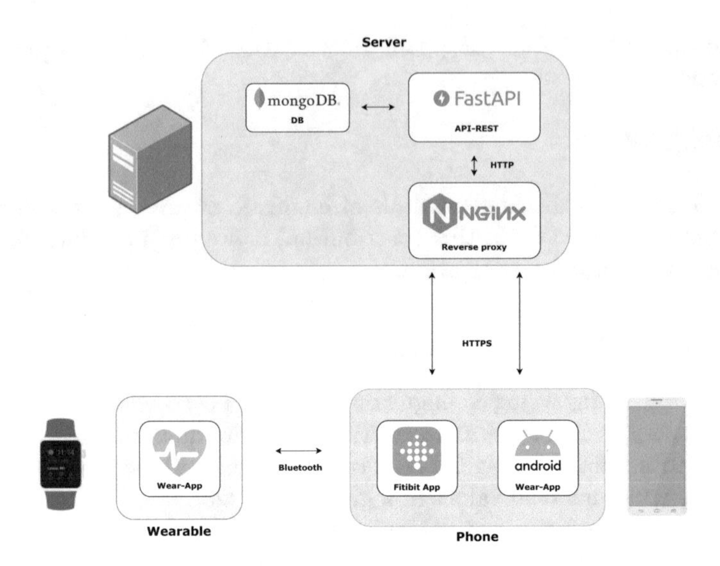

Fig. 1. Architecture of the system.

3.1 Server

This layer in the proposed architecture consists of three main elements: a reverse proxy, a REST API and a database. These elements play key roles in processing, storing and accessing data collected from wearable devices.

Reverse Proxy. The reverse proxy acts as an intermediary between clients and the server, optimizing inbound and outbound requests, enhancing performance, security, and scalability. In this architecture, it serves as a TLS termination proxy, ensuring security by managing TLS protocol termination for communications. This focus on TLS versions 1.2 and 1.3 and strong encryption algorithms enhances security while limiting backward compatibility with legacy clients. The reverse proxy also implements measures against attacks, like denial-of-service (DoS) attacks, to safeguard data and maintain high data availability.

Database. The decision to adopt a NoSQL database, specifically MongoDB, in the proposed architecture is driven by the need for flexibility and compatibility with various data-generating devices, not limited solely to Fitbit devices. MongoDB's document-based data model, utilising JSON documents for storage, facilitates efficient data management. This choice enables the system to handle heterogeneous data from diverse devices effectively [3].

API-REST. This component performs a fundamental role in the communication between the different clients, allowing data to be exchanged efficiently and securely. The REST API [10] defines a set of endpoints that represent different operations and resources available in the system. This set of resources and operations allows the different applications that make up the architecture to exchange the information needed to operate. The API has been developed using FastAPI[1], a web framework for the development of high performance APIs in Python. Among the advantages of using this framework are the ability to produce interactive documentation through the use of Swagger[2] and the ability to perform automatic validation of data types through Pydantic[3].

3.2 Devices

This layer is made up of the different applications and devices that are responsible for providing and managing the architecture's data.

Wearable. Fitbit Versa 3 offers the possibility of developing specific software, which makes it possible to customise the types of data collected as well as their frequency. In this sense, the information retrieved from the device corresponds to heart rate and acceleration, whose data collection frequencies have been set at 1 Hz and 50 Hz, respectively. The information obtained from the device is sent via customize to a service running within Fitbit's proprietary application on the mobile device. From this service the information is sent to the server via the previously mentioned API.

Phone. Within the applications that have been developed, there is one that has been designed for mobile devices, Fig. 2. The purpose of this application is to ease the management of sessions or data linked to the user's account. This application also allows the visualization of data streams and protoforms related to these streams, allowing their analysis in an intuitive way. As noted above, given that data cannot be retrieved directly from the Fitbit application, this application connects directly to the API that has been developed to obtain the data linked to the different users. The application performs data persistence, therefore when it synchronises with the server, it will download those sessions that are not available on the device. At the same time, in case of data loss from the server, it will upload those sessions that are not available on the server.

4 Proposed Linguistic Protoforms

In this section presents the formal representation of sensor streams and the linguistic protoforms used for describing the data streams of wearable devices.

[1] https://fastapi.tiangolo.com/.
[2] https://swagger.io/.
[3] https://docs.pydantic.dev/latest/.

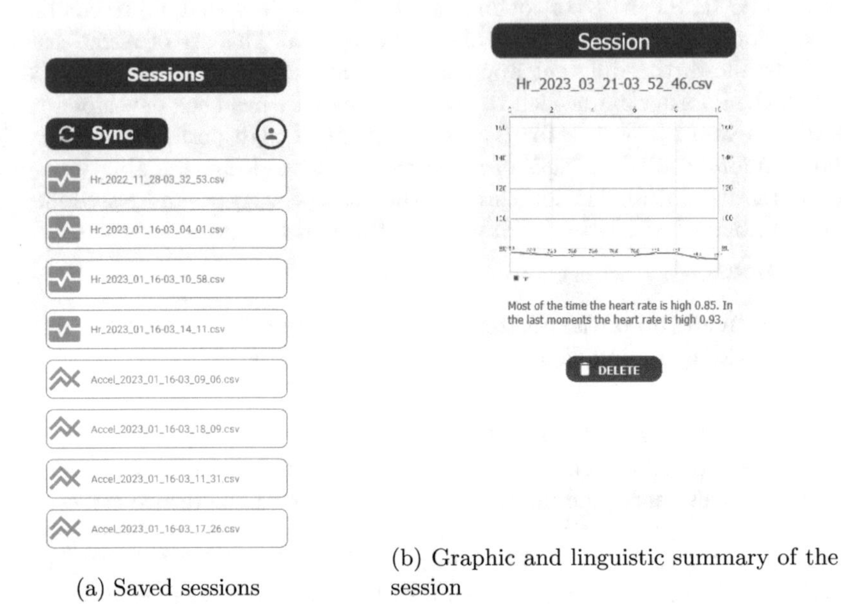

(a) Saved sessions

(b) Graphic and linguistic summary of the session

Fig. 2. Mobile application developed

4.1 Linguistic Expressions for Sensor Streams

The sensor stream s^j consists of values $s^j = \{m_i^j\}$, where $m_i^j = \{v_i^j, t_i^j\}$ represents the value and timestamp for sensor j at time t_i.

The protoform has the structure: $Q_k\ V_r\ T_j$, where V_r is a crisp term linked to event r, and T_j is a Fuzzy Temporal Window (FTW) aggregating event V_r. FTWs are defined by the temporal distance $\Delta t_i = t^* - t_i$ using membership function $\mu_{T_j(\Delta t_i)}$.

An aggregation function of V_r over T_j computes a unique degree representing the occurrence of event V_r within the temporal window T_j. The subsequent t-norm and t-conorm operators have been established to combine a linguistic term and a temporal window:

$$V_r \cap T_j(\bar{s_i^l}) = V_r(s_i^l) \cap T_j(\Delta t_i) \in [0,1]$$

$$V_r \cup T_j(\bar{s_i^l}) = \bigcup_{s_i^l \in S^l} V_r \cap T_j$$

Q_k serves as a fuzzy quantifier k responsible for filtering and transforming the aggregation degree. It applies a transformation $\mu_{Q_k} : [0,1] \rightarrow [0,1]$ to the aggregated degree.

4.2 Linguistic Protoform Based on Heart Rate

As previously mentioned, among the different values that the wearable device collects is heart rate. We propose four fuzzy terms: *low, moderate, high, very high.* The following fuzzy sets are defined using trapezoidal and triangular membership functions for heart rate, expressed in beats per minute (bpm), Fig. 3.
$\mu HRL(vhr)$: *TS(0,0,50,60)* $\mu HRM(vhr)t$: *T(50,80,100)* $\mu HRH(vhr)$: *T(80, 120,140)* $\mu HRVH(vhr)$: *TS(120,140,200,200)*

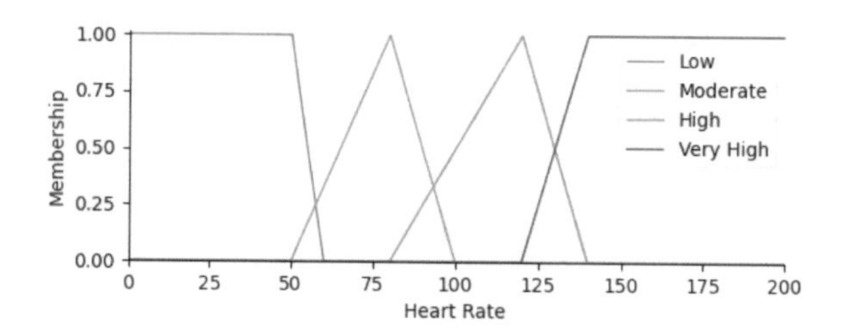

Fig. 3. Membership functions of the fuzzy terms of variable heart rate.

In addition, three fuzzy quantifiers, *part of the time*, *half of the time* and *most of the time*, defined by Gaussian and sigmoid membership functions, Fig. 4, are define
$PartOfTheTime(\Delta t)$: *S(50, -0.1)* $HalfOfTheTime(\Delta t)$: *G(50, 10)* $MostOf TheTime(\Delta t)$: *S(50, 0.1)*

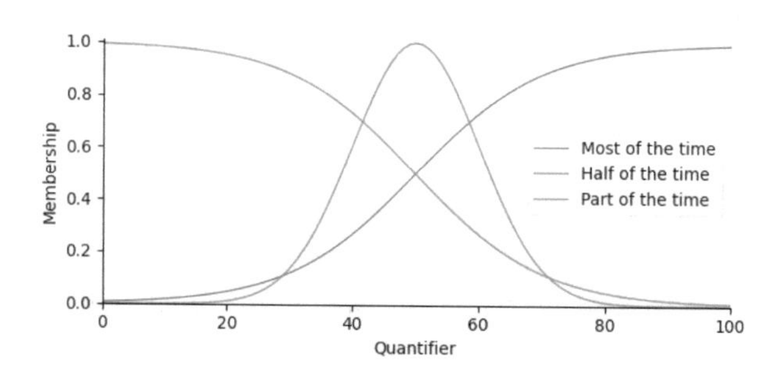

Fig. 4. Membership functions of the fuzzy quantifiers.

Based on the fuzzy sets and fuzzy quantifiers defined above, linguistic protoforms can be constructed as follows:

- $Protoform_1$:*Most of the time the heart rate is moderate*

This type of protoform aims to express concisely in natural language the general behavior of the heart rate during the development of different exercises or sports activities (Fig. 5).

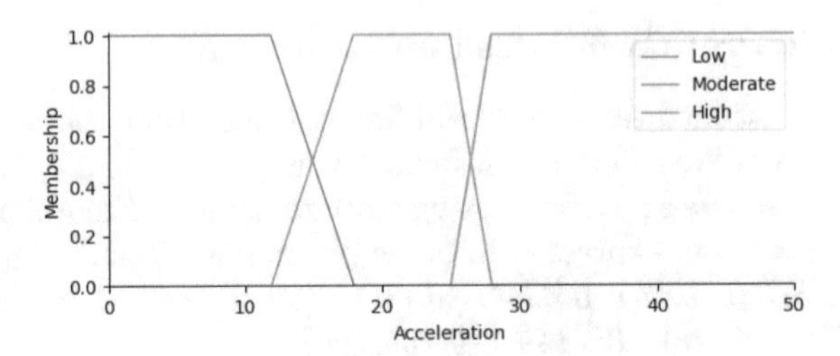

Fig. 5. Membership functions of the fuzzy terms of variable acceleration.

Finally, a fuzzy temporal window, *at the last moments*, is defined by means of a sigmoidal membership function.
$InTheLastMoment(\Delta t)$: *S(10, 0.6)*

This allows the creation of linguistic protoforms such as:

- $Protoform_1$:*In the last moments, the heart rate is very high.*

These protoforms are particularly useful, since during the final stages of certain exercises, the intensity increases, which allows us to make a comparison with the rest of the exercise.

4.3 Linguistic Protoform Based on Aceleration

Three fuzzy terms, *low, moderate and high*, which are defined by trapezoidal membership functions, are proposed for acceleration analysis. $\mu AL(va)$: *TS(0,0,12,18)* $\mu AM(va)t$: *TS(12,18,15,28)* $\mu AH(va)$: *TS(25,28,50,50)*

Combined with the fuzzy quantifiers defined above, linguistic protoforms of the following type are formed:

- $Protoform_1$:*Part of the time acceleration is high*

5 Case Study

In this section a case study composed of different scenarios will be presented, whose objective is to evaluate the proposed architecture. For this purpose, a series of exercises are proposed, where a data collection frequency of 1 sample per second for heart rate and 50 samples per second for acceleration will be set.

5.1 First Scenario Based on Relaxation Exercises

The first scenario of the case study involves performing relaxation exercises, which involve alternating between different postures.

Acceleration variations can be noted when changing posture, as shown in Fig. 6, while heart rate is maintained at normal levels.

As for the linguistic protoforms, the data obtained for heart rate are as follows: Most of the time the heart rate is normal 0.99. In the last moments the heart rate is normal 0.5. While for the acceleration it has been obtained: Most of the time the acceleration is low 0.99.

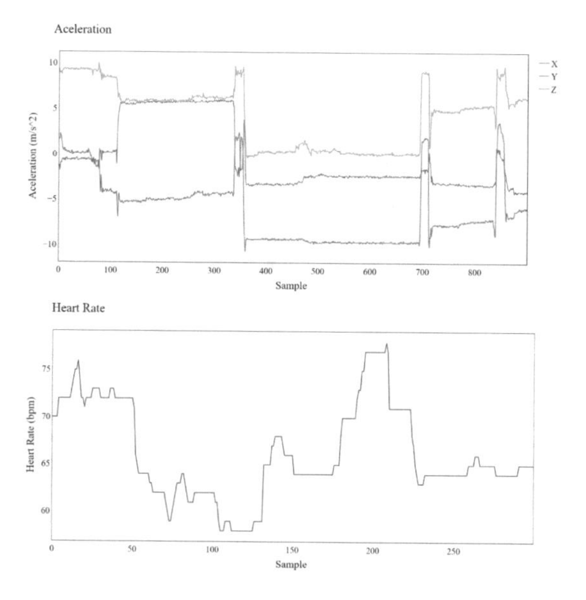

Fig. 6. Data streams of the first scenario.

5.2 Second Scenario Based on Moderate-Intensity Exercise

The second scenario of the case study consists of moderate-intensity exercise, specifically focused on bodybuilding.

Figure 7 shows discernible patterns in acceleration, which are related to bodybuilding exercises. In addition, with regard to heart rate, its evolution can be observed as the different exercises are carried out.

In relation to the protoforms, the following results are obtained: Half of the time the heart rate is high 0.73. In the last moments the heart rate is very high 1.0 and Most of the time the acceleration is low 0.99. It is interesting to note that, although the acceleration is higher compared to the previous case, it is still classified as low, suggesting the need to adjust the membership functions. However, as for the heart rate, we find useful information, such as the fact that at the end of the exercise, the heart rate is very high.

5.3 Third Scenario Based on High Intensity Exercises

The third scenario of the case study involves the performance of high intensity exercises. In this case, the exercise known as "burpees" has been selected as the specific exercise to be performed.

During this scenario, a series of four repetitions of burpees is performed. A burpee is an exercise that involves a combination of movements such as a push-up, a vertical jump and a squat. This exercise is characterized by its high cardiovascular and muscular demand, making it an intense activity.

Based on Fig. 8, both high accelerations and high heart rate can be observed.

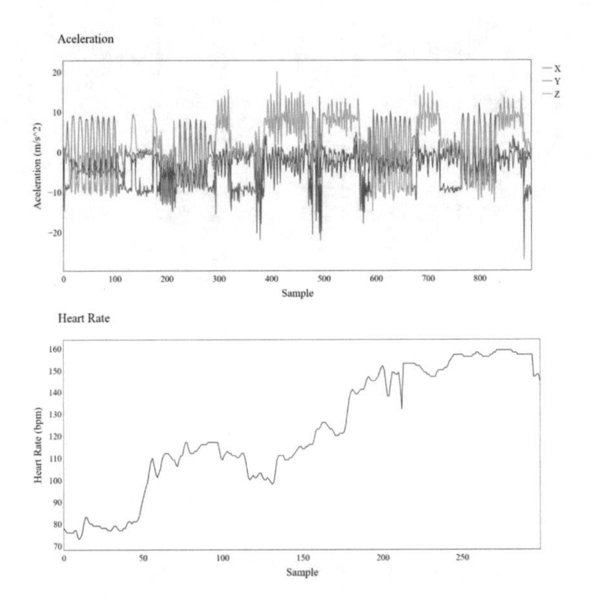

Fig. 7. Data streams of the second scenario.

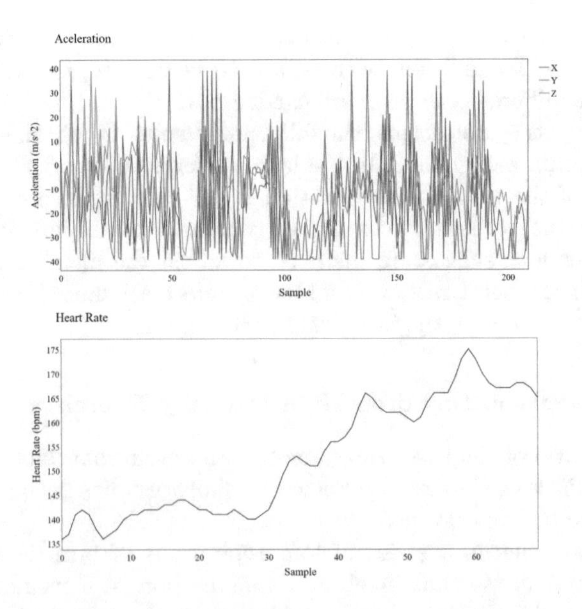

Fig. 8. Data streams of the third scenario.

Regarding the obtained linguistic protoforms, we have the following results: Most of the time the acceleration is high 0.71 and Most of the time the heart rate is very high 0.99. In the last moments the heart rate is very high 1.0.

6 Conclusions

In this contribution, a new advanced architecture for wearable device data collection and processing has been proposed. Compared to other commercial alternatives, this architecture provides raw data, allowing the most appropriate processing to be applied to each user's individual requirements.

Furthermore, an additional approach is provided by the application of linguistic summaries based on the use of fuzzy logic. These linguistic summaries ease the analysis of the collected data by providing an intuitive and understandable representation of the data, thus facilitating interpretation and informed decision making.

In addition, a comprehensive case study has been developed, comprising various scenarios, to evaluate the proposed architecture thoroughly. This case study served as a valuable tool for assessing the effectiveness and practicality of the proposed architecture.

In future work, we intend to further improve this architecture by incorporating more complex and personalized fuzzy protoforms, so that they can describe more accurately and in more detail the data streams.

References

1. Albín-Rodríguez, A.P., De-La-Fuente-Robles, Y.M., López-Ruiz, J.L., Verdejo-Espinosa, Á., Espinilla Estévez, M.: UJAmI Location: a fuzzy indoor location system for the elderly. Int. J. Environ. Res. Public Health **18**(16), 8326 (2021)
2. Albín-Rodríguez, A.P., Ricoy-Cano, A.J., de-la Fuente-Robles, Y.M., Espinilla-Estévez, M.: Fuzzy protoform for hyperactive behaviour detection based on commercial devices. Int. J. Environ. Res. Public Health **17**(18), 6752 (2020)
3. Bradshaw, S., Brazil, E., Chodorow, K.: MongoDB: the definitive guide: powerful and scalable data storage. O'Reilly Media (2019)
4. Díaz, D., Medina, J., Montoro, A., López, J.L., Espinilla, M.: Linguistic summaries for dwellings energy poverty monitoring. In: International Conference on Ubiquitous Computing and Ambient Intelligence, pp. 693–704. Springer, Cham (2022). https://doi.org/10.1007/978-3-031-21333-5_69
5. Espinilla, M., Medina, J., García-Fernández, Á.L., Campaña, S., Londoño, J.: Fuzzy intelligent system for patients with preeclampsia in wearable devices. Mob. Inf. Syst. **2017**, 1–10 (2017). https://doi.org/10.1155/2017/7838464
6. Lou, Z., Wang, L., Jiang, K., Wei, Z., Shen, G.: Reviews of wearable healthcare systems: materials, devices and system integration. Mater. Sci. Eng. R. Rep. **140**, 100523 (2020)
7. Lu, L., et al.: Wearable health devices in health care: narrative systematic review. JMIR Mhealth Uhealth **8**(11), e18907 (2020)

8. Marín, N., Sánchez, D.: On generating linguistic descriptions of time series. Fuzzy Sets Syst. **285**, 6–30 (2016). https://doi.org/10.1016/j.fss.2015.04.014, https://www.sciencedirect.com/science/article/pii/S0165011415002110, special Issue on Linguistic Description of Time Series
9. Martinez-Cruz, C., Rueda, A.J., Popescu, M., Keller, J.M.: New linguistic description approach for time series and its application to bed restlessness monitoring for eldercare. IEEE Trans. Fuzzy Syst. **30**(4), 1048–1059 (2022). https://doi.org/10.1109/tfuzz.2021.3052107
10. Masse, M.: REST API design rulebook: designing consistent RESTful web service interfaces. O'Reilly Media, Inc. (2011)
11. Miotto, R., Wang, F., Wang, S., Jiang, X., Dudley, J.T.: Deep learning for healthcare: review, opportunities and challenges. Brief. Bioinform. **19**(6), 1236–1246 (2018)
12. Nahmias, S.: Fuzzy variables. Fuzzy Sets Syst. **1**(2), 97–110 (1978). https://doi.org/10.1016/0165-0114(78)90011-8
13. Oresko, J.J., et al.: A wearable smartphone-based platform for real-time cardiovascular disease detection via electrocardiogram processing. IEEE Trans. Inf Technol. Biomed. **14**(3), 734–740 (2010)
14. Parkka, J., Ermes, M., Korpipaa, P., Mantyjarvi, J., Peltola, J., Korhonen, I.: Activity classification using realistic data from wearable sensors. IEEE Trans. Inf Technol. Biomed. **10**(1), 119–128 (2006)
15. Patel, S., et al.: Monitoring motor fluctuations in patients with Parkinson's disease using wearable sensors. IEEE Trans. Inf Technol. Biomed. **13**(6), 864–873 (2009)
16. Peláez-Aguilera, M.D., Espinilla, M., Olmo, M.R.F., Medina, J.: Fuzzy linguistic protoforms to summarize heart rate streams of patients with ischemic heart disease. Complexity **2019**, 1–11 (2019). https://doi.org/10.1155/2019/2694126
17. Ravì, D., et al.: Deep learning for health informatics. IEEE J. Biomed. Health Inform. **21**(1), 4–21 (2016)
18. Son, D., et al.: Multifunctional wearable devices for diagnosis and therapy of movement disorders. Nat. Nanotechnol. **9**(5), 397–404 (2014)
19. Zadeh, L.: The concept of a linguistic variable and its application to approximate reasoning-III. Inf. Sci. **9**(1), 43–80 (1975). https://doi.org/10.1016/0020-0255(75)90017-1
20. Zadeh, L.: Fuzzy logic. Computer **21**(4), 83–93 (1988). https://doi.org/10.1109/2.53

Face Recognition and Actuation to Promote Smart Home Security

Jayakrishnan Harisadethan Nair, Paul McCullagh(✉), and Ian Cleland

Ulster University, Belfast, Northern Ireland, UK
pj.mccullagh@ulster.ac.uk

Abstract. Internet of Things (IoT) provides the opportunity for remote sensing in a smart home. This can benefit home security. A case study was devised to remotely sense, inform and actuate. The study developed a surveillance system using Raspberry Pi and camera to detect and recognize a door caller's face from a live video stream and compare with an existing dataset to verify whether the person was authorized. If not authorized, the remote homeowner can be alerted via an email complete with an image of the caller. Automated and user actions can be performed depending on the type of visitor. RaspController, Google Assistant and VNC were used for actuation, to perform actions like playing random music, triggering voice conversations or turning on an alarm. The recognition system was developed in Python using OpenCV and face recognition, using pre trained models for Hog detection and Haar Cascade classification. The accuracy of the detection was 66.7% for a small cohort of seven authorized people and an unauthorized class of five people. For actuation, interoperability with many smart home devices can be controlled from the smart home network, but with only limited functionality from a remote network.

Keywords: Home Security · IoT · Raspberry Pi · Face recognition · OpenCV · Google Assistant

1 Introduction

Internet of Things (IoT) refers to an ecosystem where a vast number of devices are interconnected to gather or share data via the Internet in real time [1]. Due to the rapid growth of smart home devices and the interconnection between the devices using IoT, smart surveillance systems can be implemented to protect the home from unauthorized persons or intruders [2]. The use of cameras for monitoring activity at front doors has become commonplace, with solutions from companies such as Google and Amazon. Face recognition software has been used by researchers to extend the functionality of such systems [3]. However, this requires a bespoke solution for sensing and recognition. Subsequent actuation is possible with the installed smart home devices (e.g., speakers, lights, alarms; from a large number of vendors) and this allows us to address the current state of interoperability of IoT.

The Raspberry Pi (RPi) [4] is a miniature sized single board computer which can be used for various IoT applications due to its low power consumption and versatility. RPi

J. Bravo and G. Urzáiz (Eds.): UCAmI 2023, LNNS 841, pp. 151–158, 2023.
https://doi.org/10.1007/978-3-031-48590-9_14

can be used for implementing bespoke surveillance systems with the help of suitable libraries supported by OpenCV [5] and bespoke development in Python [6].

Amir et al. [7] suggested a home surveillance system where motion was detected by change in frames of an RPi camera. The authors implemented face recognition using a local binary pattern (LBP) algorithm [8], which facilitated authorization. If the image stored in the memory matched with the live image, approval would be given to the visitor; alternatively, the software notified the owner through an SMS about an unknown visitor. Pryai et al. [9] reported an RPi doorbell system with actuation. When a visitor came to the door and pressed the doorbell, the camera was triggered and captured the visitor's face. If the visitor was authorized (i.e., preregistered in a database), the door opened automatically using actuation (RPi's GPIO used as a trigger). If an unauthorized person was at the door, an image of the visitor was sent to the homeowner's mail. Similarly, Jyothi et al. [10] used RPi and RPi camera to identify movement. The motion detection algorithm subsequently triggered alert messages.

Awais et al. [11] implemented a facial detection and recognition system. The classification was carried out by a Multilayer Feed Forward Neural Network [12]. Sajjad et al. [13] proposed recognition with the Viola Jones face detector using Orient FAST and Rotated BRIEF (ORB) points markers. Due to the limited computational capacity of the RPi, a Support Vector Machine (SVM) was used for training and recognition was carried out in the cloud. Only the extracted features of the image were uploaded to the cloud server to minimize the transmission time and bandwidth.

IoT technology can be used to assist vulnerable people who may wish to restrict access or to infer that multiple occupants are present if a caller is at the door. In this paper, we extend the work on sensing, face detection and authorization and assess the tools that can be used to interoperate with existing smart home devices in a Google Home.

The paper is organized as follows. Section 2 provides implementation details, specifically the system design and the creation of a dataset to assess the accuracy of the face recognition component. Actuation possibilities are then evaluated. Section 3 provides results of the testing component. Discussion and conclusion are provided in Sect. 4.

2 Implementation

2.1 System Design

The system was based on the RPi with camera for image acquisition. Face detection for the smart home visitor was implemented by pre-trained models in OpenCV. The simple mail transport protocol (SMTP) library supported by Python was selected for email alert to the homeowner. Actuations were to be controlled by RaspController, if on the same network as the Rpi (home network), and virtual network computing (VNC) [14], if on a different remote network. Google Assistant API was selected for voice commands to initiate responses.

The RPi security system was developed using the `pypi` library for face recognition [15] and the `imutils` library for image processing [16]. A dataset was captured using the Pi camera and stored in the memory as a 'whitelist' of authorized persons. Face detection was carried out by pre-trained models provided by `OpenCV`, namely Histogram of Oriented Gradient (Hog) classifier [17] for registration and Haar Cascade classifier [18] for identification. The features from the images were encoded as 128-dimensions and the embeddings stored in a Python 'pickle' file [19].

Live images/frames from the video streaming were subsequently captured and the features extracted by the Haar classifier to 128-dimension encodings. The features were then compared with the embeddings stored in the pickle file and the appropriate face was identified by a voting procedure, in real-time. After recognizing a face, the captured image was embedded in an email to notify the homeowner (who may be remote) about the visitor. If the visitor was unauthorized a buzzer alarm was automatically activated at the home for deterrence. Based on the email feedback, the homeowner was then capable of smart home actuation; remotely accessing the RPi to perform actions such as playing random songs or recorded conversation, reactivating the buzzer or triggering the RPi Google Assistant.

2.2 Creation of Dataset, Face Recognition and Actuation

A python file named `Capture.Py` was used for capturing JPEG images for the 'whitelist' of authorized persons. Images were stored in a directory named after the registered user. This dataset was then used for encoding face embeddings using appropriate pre-trained classifiers provided by OpenCV package. Images of well-known personalities were used for the dataset to expedite development and ameliorate any privacy concerns.

Captured images were preprocessed by resizing frames and conversion to grayscale to reduce computational load and increase accuracy. The Hog classifier's `Detect_multiscale()` method was used to perform the face detection. Bounding boxes detected faces within the image which were stored in memory for further computation. Using the `Face_encodings()` method, 128-dimension embeddings of the face were computed. After processing all the images, the serialized embeddings were written into a pickle file for subsequent face recognition.

The Harr cascade classifier was used for detecting face locations from live video streaming, as it has less computational requirements. The `Videostream` class from `Imutils` package was used to initialize the video stream using RPi camera module. A loop was used to continuously acquire frames from the live stream. To increase the processing speed the frame was resized to 320 by 320 pixels. The `Face_locations()` method was used to find the bounding boxes in the frame. When the face was detected, the image was encoded into 128-d embeddings with `Face_encodings()` method. Stored embeddings of the dataset in the pickle file were compared with the data from live stream image. Recognition used a voting method and the image with highest number of votes was considered as the recognized face. If the match was found the name was appended to the screen for subsequent feedback.

The next stage after face recognition was to alert the homeowner about the identity of the visitor. The captured image along with a subject line to notify the type of visitor was sent by email from RPi. This utilized a Gmail account and the `Smtplib` library in Python.

The Google Assistant API available in the google console actions dashboard [20] was used to implement google assistant in RPi. The following steps were required.

step1: a new project was created for RPi Google Assistant. In the google developers console, the project was selected and the Google Assistant API enabled. The same `Model_id` was then registered in RPi.

step2: OAuth 2.0 credentials were downloaded. This was a JSON file which was stored in the same folder as the google assistant software development kit. It was necessary to configure the OAuth consent screen [21]. The user type was set to external so that any person with valid email account can access the project.

step3: It was necessary to enable access to the web and app activity, location history, device information and voice and audio activity.

step4: Before activating the Google Assistant in RPi, the email id used for creating the model was added to the test user. Otherwise, the authorization failed because of the permission issues.

Raspcontroller is an android application developed to manage and control the RPi with the use of a smart phone [22]. It eliminates the need for direct connection of a monitor, keyboard, and mouse to the RPi. Commonly used scripts can be saved to increase usability. However, a limitation is that the phone must be connected to the same network as the RPi.

Virtual Computer Networking (VNC) can be used to access the Raspberry Pi from a remote network. The IP address of the Raspberry Pi was essential to establish the connection. This can be identified by running the command `Hostname -I` in the console. Authentication using username and password was required to establish the connection between the device and the Raspberry Pi.

3 Results

3.1 Performance of the Classifier

The RPI camera module was used for live video capture. Using the Haar Cascade Classifier, faces were identified; bounding boxes provided feedback and multiple faces in a frame could be identified. 128-d embeddings of the detected faces were extracted. Figure 1 shows the output obtained for selected authorized and unauthorized persons. The confusion matrix for the predicted result is shown in Table 1. The accuracy of the detection was 66.7% for a cohort of seven authorized people (six of whom are shown) and an unauthorized class.

Accuracy can be increased by adding to the numbers of users (authorized and unauthorized). The number of users is small, as we are demonstrating proof of concept of the actuation component using pre-tested recognition algorithms. Additional people successfully tested the recognition, but their data were not included due to ethical concerns regarding privacy of storing images.

3.2 Actuation and Limitations on Interoperability

After recognizing a face some automatic actions can be performed. The aim was to give the impression of multiple occupancy within the home with spurious ongoing audible activity. If the visitor is one of the authorized people, an LED will be turned on automatically and will turn off after 5 s. Similarly, when an unauthorized person is recognized,

Fig. 1. Facial recognition of 'whitelist' (left) and unauthorized people (right).

Table 1. Confusion matrix for dataset with seven authorized people (six shown in Fig. 1) and an unauthorized group of five people (four shown in Fig. 1).

	Jay	Paul	Tom	Brad	Ang	Sand	Seli	Unau
Jay	1	0	0	0	0	0	0	0
Paul	0	1	0	0	0	0	0	0
Tom	0	1	1	0	0	0	0	0
Brad	0	0	0	1	0	0	0	1
Ang	0	0	0	0	1	0	0	0
Sand	0	0	0	0	0	1	0	0
Seli	0	0	0	0	0	0	1	0
Unau	0	0	0	0	0	0	0	5

a buzzer/alarm will be activated automatically and will shut down after 5 s. These actuations could be used within the home for feedback and to deter unwanted visitors. An email with an attachment of the captured image is sent to the homeowner. When the alert is received, the owner can remotely access the RPi. There are two scenarios:

Scenario 1: The owner was connected to the home network. The live camera feed was monitored using RaspController to verify the authenticity of the alert received. The GPIO pin was used to turn the LED on and activate the buzzer. The Google Assistant 'Hello/Hey Google' activation command was used. In the experiment we asked Google Assistant for the current weather information and subsequently asked Google Assistant to tell some jokes. Functionality that required more memory, such as connection to Spotify was not possible due to the constrained nature of the RPi device.

Scenario 2: The owner was connected to a different network (i.e., remote, across the internet). Connectivity was possible to the RPi requires using VNC and a list of random songs stored on the RPi was played, giving the impression of occupancy. Recorded conversations were also played using VLC media player.

4 Discussion and Conclusions

Our aim was to assess interoperability in Smart Home IoT, within the Google Home ecosystem, which has access to hundreds of devices. The RPi hosted Google Assistant carried out 'lightweight' commands like telling jokes and providing current weather status. VNC was used to connect to the RPi from outside the home network and used the console to play random songs and recorded conversations to make any unauthorized visitor believe someone was inside the house. This relied on data stored within the RPi file system.

Since RPi has limited processing power and memory, it is difficult to train complex deep learning models on this device. If complex models need to be implemented in RPi, an efficient method is to train the model on a more powerful machine or cloud-based platforms and then transfer the trained model back to the RPi. To reduce the time for training and testing of datasets, the solution opted for was to use pre-trained models supported by the OpenCV package. Hog and Haar Cascade classifiers were utilised. The technical comparison of both models is shown in Table 2.

Table 2. Technical Comparison between Hog and Haar Cascade Classifiers.

Pre-trained Hog classifier	Pre-trained Haar classifier
Detects objects by analyzing the distribution of gradient orientations	Detects objects by analyzing patterns of intensity variations
Suitable for detecting various objects such as humans, vehicles	Used for detecting faces and other specific objects
Capture the shape and edge information	Capture simple patterns and textures in objects
Higher accuracy in object detection	Faster processing speed for real-time tasks
Requires more computational resources due to more detailed feature extraction	Requires less computational resources due to simpler features
Can handle variations in pose, scale, and orientation of objects	Less robust to variations

The Hog classifier was used for detecting faces for the registration dataset. The dataset consisted of several images of the same person having different characteristics including pose variation and lighting. It was required to detect faces from those images with high accuracy. Alternatively, when real time face detection was required, a classifier with better computational efficiency and high processing speed was needed. The Haar Cascade classifier was used to detect faces from live frames as it proved compatible, with appropriate accuracy, with real time detection.

When the RPi camera was used for live video capture, the main issue faced was achieving a high enough frame rate. This was due to limited processing power. The capturing of images was slow when the resolution was set to industry standard full high-definition size (e.g., 1920 by 1080 pixels). When it was reduced to approximately 320 by

320 pixels, usability increased, and the frame rate increased from 1.2 frames per second to 6.2 frames per second.

However, the full functionality of smart home control could not be achieved, as the normal actuation relies on voice commands to Google Assistant, which needed to be initiated from the home network. Whilst devices could be controlled remotely using their individual APIs, the convenience of Google home actuation was not readily exposed.

The RPi is a constrained device, and it would struggle to handle large number of faces in database and hence the pickle file approach was adopted. The classification algorithms we used worked appropriately on the RPi, but there will be reduced identification accuracy when confronted with variations in lighting, pose angles. The imaged-based authorization approach is limited, as it would be possible to fool the recognition component with a photograph of a 'white-listed' visitor. Thus, as is the current trend, multifactor authentication may be needed, e.g., with an additional voice recognition component. Indeed, if multiple people were identified in an image further verification would be required.

Several improvements are possible. Face recognition can be more efficient if complex deep learning algorithms like CNN are used [23]. It is not possible to train the model directly in RPi, so the training model must be performed on a high-end system with model parameters transferred to the RPi. Additionally, it would be possible to synthesize speech to activate Google Assistant API to control the plethora of devices registered to the home. A further improvement would be to add text messaging in addition to email functionality for expedited feedback.

In conclusion, we found that interoperability with Google Home is good when devices are registered to the home. However, full control of Google Assistant requires local network access, with more limited functionality available remotely by VNC.

References

1. Rayes, A., Salam, S.: Internet of Things: From Hype to Reality: The Road to Digitization, 2nd edn. Springer, Cham (2019). https://doi.org/10.1007/978-3-319-99516-8
2. Akter, S., Sima, R.A., Ullah, M.S., Hossain, S.A.: Smart security surveillance using IoT. In: Proceedings of the 2018 International Conference on Reliability, Infocom Technologies and Optimization (2018)
3. Lalitha, R., Kavitha, K., Rao, N., Mounika, G., Sandhya, V.: Smart surveillance with smart doorbell. Int. J. Innov. Technol. Explor. Eng. **8**(8), 1841 (2019)
4. Raspberry Pi. http://www.raspberrypi.org/. Accessed 10 July 2023
5. OpenCV Documentation. https://docs.opencv.org/4.x/d9/df8/tutorial_root.html. Accessed10 July 2023
6. Python. https://www.python.org/. Accessed10 July 2023
7. Aamir, N., Mohamed, S.: An internet of things approach for motion detection using raspberry pi. In: Proceedings of the International Conference on Intelligent Computing and Internet of Things (2015)
8. Ojala, T., Pietikainen, M., Maenpaa, T.: Multiresolution gray-scale and rotation invariant texture classification with local binary patterns. IEEE Trans. Pattern Anal. Mach. Intell. **24**(7), 971–987 (2002). https://doi.org/10.1109/TPAMI.2002.1017623
9. Priya, P., Viraj, M.: Smart motion detection system using Raspberry Pi. Int J. Appl Inform. Syst. **10**(4), 37–40 (2016)

10. Jyothi, S., Vijaya, V.: Design and implementation of real-time security surveillance system using IoT. Int. Conf. Commun. Electron. Syst. (2016). https://doi.org/10.1109/CESYS.2016.7890003
11. Awais, M., Iqbal, M., Ahmad, O., Alassafim M. , Alghamdi, R., Basheri,, M., Waqas, M., Real-Time surveillance through face recognition using HOG and feedforward neural networks, IEEE Access, **vol**. 7, (2019)
12. Himavathi, S., Anitha, D., Muthuramalingam, A.: Feedforward neural network implementation in FPGA using layer multiplexing for effective resource utilization. IEEE Trans. Neural Networks **18**(3), 880–888 (2007)
13. Sajjad, M., et al.: Raspberry Pi assisted face recognition framework for enhanced law-enforcement services in smart cities. Future Gener. Comput Syst **108**, 995–1007 (2020). https://doi.org/10.1016/j.future.2017.11.013
14. Connecting to Raspberry Pi via VNC. https://raspberrypiguide.github.io/networking/connecting-via-VNC. Accessed 10 July 2023
15. Python library for face recognition. https://pypi.org/project/face-recognition. Accessed 10 July 2023
16. Python library for image processing. https://pypi.org/project/imutils. Accessed 10 July 2023
17. Mallick, S., Histogram of Oriented Gradients explained using OpenCV, Learn OpenCV. https://www.learnopencv.com/histogram-of-oriented-gradients/. Accessed 10 July 2023
18. OpenCV Cascade Classifier Tutorial. https://docs.opencv.org/3.4/db/d28/tutorial_cascade_classifier.html. Accessed 10 July 2023
19. Python pickle module - Saving Objects via Serialization. https://pythonprogramming.net/python-pickle-module-save-objects-serialization. Accessed 10 July 2023
20. Google Assistant SDK Documentation. https://developers.google.com/assistant/sdk/guides/library/python. Accessed 10 July 2023
21. PiAsst - APIs & Services - Google Cloud Console [Online]. https://console.cloud.google.com/apis/library/piasst. Accessed 10 July 2023
22. RasPiConnect - Use Your Phone to Control Your Raspberry Pi. https://core-electronics.com.au/tutorials/raspicontroller-use-your-phone-to-control-your-raspberry-pi.html. Accessed 10 July 2023
23. Paul, S., Acharya, S.: A comparative study on facial recognition algorithms (December 21, 2020). e-journal - First Pan IIT International Management Conference (2018). https://ssrn.com/abstract=3753064https://doi.org/10.2139/ssrn.3753064

IoT-Watcher: A Climate Monitoring System for IT Infrastructures Based on IoT and Open-Source Technologies

Alejandro Mosteiro[1(✉)], Carlos Dafonte[2], Ángel Gómez[2], Daniel Boubeta[2], and Manuel G. Penedo[2]

[1] CITIC, Universidade da Coruña, Campus de Elviña s/n, 15701 A Coruña, Spain
a.mosteiro@udc.es

[2] CITIC - Department of Computer Science and IT, Universidade da Coruña, Campus de Elviña s/n, 15071 A Coruña, Spain
https://citic.udc.es

Abstract. In this paper we describe a monitoring system for IT environments such as Data Centers or other IT infrastructures, completely based on low cost open-hardware and open-source technologies. We have designed an ad-hoc PCB based on the ESP-8266 microcontroller integrating all the needed sensors. All data collected using MQTT communications is then sent to a Zabbix server where we integrate environmental data with networks and systems monitoring.

The aim of this work is to provide Data Centers with an open-source monitoring system alternative, giving administrators a powerful tool that integrates Arduino-compatible open-hardware sensors integrated in a Zabbix environment. The system is based on IoT protocols and standards, with the objective of it being as flexible and modular as possible. Zabbix can monitor multiple parameters on a network and compute servers and can be configured to send alerts or other events and can also work as a reporting and data visualization tool. Integrating IoT devices into a Zabbix server offers the possibility of monitoring virtually any parameter administrators may think of.

The hardware devices designed have been tested and we show data comparing our system with data obtained from the actual CITIC Data Center's monitoring system, a proprietary solution, so we can offer a correlation between these two systems.

Keywords: Monitoring · Zabbix · Data Center · Sensors · Arduino · Internet of Things

J. Bravo and G. Urzáiz (Eds.): UCAmI 2023, LNNS 841, pp. 159–165, 2023.
https://doi.org/10.1007/978-3-031-48590-9_15

1 Introduction

CITIC (Center for Information and Communications Technology Research) of the University of A Coruña is a research center focused on five main research subjects: Artificial Intelligence, Data Science, Cloud Computing, Intelligent Networks and Services and Cybersecurity. It has a Data Center that offers its services to over 200 researchers working in the research areas mentioned. The data center has a 2-row hot-aisle layout called "the Cube" and it hosts around 300 pieces of hardware at the time of writing this document.

The Cube has a proprietary monitoring system developed by the manufacturer of the solution that integrates data from temperature, humidity and power sensors. Each of the monitored elements has to be licensed and all sensors have to be supported by the manufacturer. The cost of this product ranges between 7,000 and 30,000 euros [1] and it has important limitations for CITIC: on the one hand, the addition of new sensors is linked to the purchase of new licenses; on the other hand, only supported sensors can be added; furthermore, it does not natively support short message notification channels such as Telegram; finally, it is not possible to monitor servers and network devices with this system.

In this paper we describe a low cost open-source and open-hardware system that can be customized and can act as a powerful tool for IT administrators, that could replace expensive alternatives as the one mentioned above.

2 Related Work

There are several works related to climate monitoring in environments other than Data Centers that refer to Arduino elements, Raspbery Pi and other low-cost components [2,3]. We have also seen academic papers try to solve problems similar to those raised in this project. Some of them describe systems based on Arduino UNO devices with ESP8266 Wi-Fi modules, others use the NodeMCU device and DHT22 sensors to obtain temperature and humidity data inside Data Centers, centralizing the data on a Zabbix server and sending it via Telegram [4,5]. These examples deal with concepts such as those proposed in this article, but do not address the use of Home Automation tools, the integration of server parameters and other Data Center elements, or the design of ad-hoc devices to capture data from sensors without having to resort to the use of breadboards.

Another example is the use of devices that integrate the ESP8266 to monitor Data Center environments with a DHT11 sensor, programming alerts and notices via SMS on a cloud platform [6]. However, in this work the use of an ad-hoc board is not proposed either, instead breadboards are used to connect the different components.

After studying the cited works on the subject, it was concluded that it would be interesting to test sensors and software traditionally used in IoT applications in a Data Center environment, designing ad-hoc boards that allow the developed solution to be more robust and then integrating this data with server and network devices monitoring as well as making sure the developed system is completely open-source.

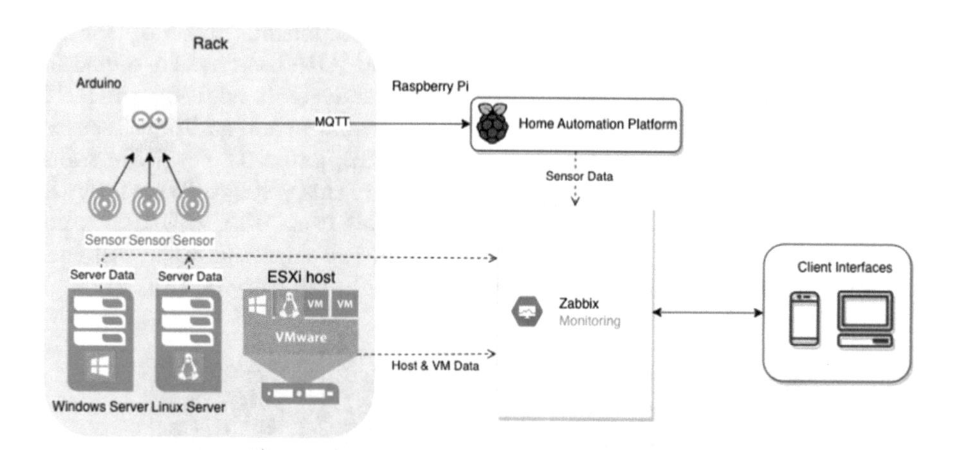

Fig. 1. Architecture Diagram

3 Architecture and Sensors

The system we propose is based on open-source technologies and tools, and is designed in such a way that each of the components that make it up can be replaced with other alternatives. These are the three main components (see Fig. 1):

- Sensors. We have designed and built a board based on ESP8266 microcontroller where data from temperature, humidity and presence sensors is collected. This board is called "SensorBoard", and could be replaced with other sensors or boards as long as the new hardware works with the MQTT topics defined on the server.
- Home Automation Platform. Data from sensors is sent to this platform, responsible for sending all information to the Zabbix server. We use Home Assistant, but it would be possible to replace this with other Home Automation software such as openHAB.
- Monitoring system. In this proposal we are using Zabbix as the main monitoring system. It may also be possible to use other monitoring systems, such as Nagios or Prometheus, after configuring them to receive MQTT messages from the topics previously defined.

3.1 Ad-Hoc Board Design

After validating the operation of all the elements on prototyping boards, a PCB was designed to integrate all the components into a more robust product. The temperature, humidity and presence sensors are connected to this PCB. An OLED screen has also been added to display the device name along with the temperature and humidity data on the device. The ESP8266 is used in standalone mode, taking the schematics for the NodeMCU development board as reference [7], so the PCB includes the passive components necessary for the microcontroller to work. Digital pins IO0, IO2 and IO16 have been assigned for humidity, temperature and presence sensors, respectively. In addition, the IO12, IO13, IO14 and IO15 pins are left available to be able to add additional sensors if desired. The board is powered by 5 V Micro USB, and a 3.3 V voltage transformer is included to power the sensors that require this voltage. To simplify the circuit and reduce costs, a chip to translate signals from USB to UART is not included, so an adapter is necessary to program the microcontroller. All code and schematics are publicly available in the official repository [8] (Fig. 2).

Board top view Board bottom view

Fig. 2. Render of the PCB and case design made with Fusion 360

4 Test and Results

In order to test the system we have placed sensors inside two racks of the Data Center. In each of those racks we installed temperature and humidity sensors close to those of the proprietary system (See Fig. 3) and a PIR sensor for presence detection.

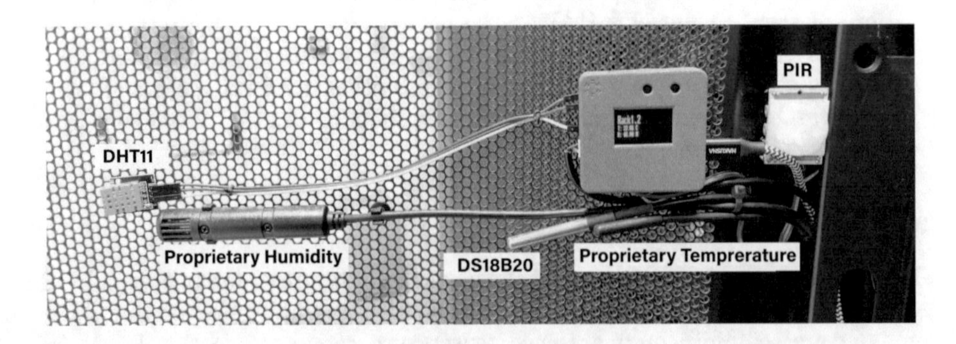

Fig. 3. Sensor installed on the Cube

We deployed a server instance by running a Home Assistant, Eclipse Mosquitto and Pi-hole instances on a Raspberry Pi 4. Our boards and the Raspberry Pi are connected over a Wi-Fi network isolated from the rest of the communications of the Data Center. In a separate virtual machine we have deployed a Zabbix server running on Ubuntu Server 20.04 LTS. This Zabbix instance monitors the sensors connected to our boards as well as physical and virtual servers inside the Data Center.

Table 1 presents a comparison of the minimum, maximum and mean temperature and humidity values for the two systems. Very similar values can be seen between both systems, presenting a greater difference in the mean of the rack 2.2 sensors. Despite this, it can be inferred from the data of both systems that for rack 1.2 there is an average temperature between 22 and 23 °C and for rack 2.2 an average temperature between 23 and 24 °C. With these results, we can say that the developed system is capable of taking measurements reliably, but it would be necessary to calibrate the sensors before placing them in the racks.

The system behaves in a similar way when it comes to the humidity data, it can be deduced that both systems measure a range of very similar values. In the case of the rack 1.2 sensors, it can be seen that the relative humidity is around 65% on average, with a difference of 1% in the minimum value and both with a maximum of 76%. In the case of the rack 2.2 there is a greater difference in the maximum value. As with the temperature data, these data seem to indicate that a more precise calibration of the sensors should translate into a greater similarity in the measurements.

Table 1. Temperature (°C) and humidity (%) data captured by both systems

	Temperature			Humidity		
	Min.	Avg.	Max.	Min.	Avg.	Max.
Rack 1.2 Ours	20.94	22.17	25.00	45.00	64.28	76.00
Rack 1.2 Proprietary	22.00	22.8	26.00	46.00	65.00	76.00
Rack 2.2 Ours	21.06	22.99	25.00	40.00	60.79	75.00
Rack 2.2 Proprietary	22.00	23.80	26.00	41.00	58.00	70.00

We have tested alarm triggers over Telegram and the system is capable of reliably sending alerts when certain events occur. For example, the system sends an alert each time someone access a monitored rack or when the established temperature or humidity thresholds are exceeded.

5 Conclusions and Future Work

We have designed a system based on low-cost open-source and open-hardware technologies, which allows each of the components that make it up to be exchanged and can be easily expanded. For this, we have designed and manufactured a functional prototype of a PCB for data capture that is capable of offering data to the system in a stable manner, through secure communications. This system is currently active in the CITIC CPD and is used for its daily operations, facilitating the management of critical assets in the center.

With the data presented in this article, it can be seen that a system with the characteristics described can be used reliably in a Data Center environment, being able to compare its operation to commercial systems. We have been able to verify, with the tests carried out, that the sensors work properly for long periods of time. However, in view of these data, we believe that it would be of great interest to calibrate the sensors, which could result in greater precision and reliability of the measurements, improving the overall functionality of the system.

It is planned to integrate fault detection modules based on historical data using artificial intelligence techniques, as well as introduce improvements that make it easier to control the use of resources to achieve energy savings, turning off servers or idle virtual machines. Finally, the use of more powerful microcontrollers, such as the ESP32, has also been considered for subsequent iterations of the project, so that a greater number of sensors can be connected to each device.

Acknowledgment. The realization of this work has been possible thanks to the collaboration of the CITIC of the UDC.

References

1. Data center expert pricing. https://www.se.com/es/es/product-range/61851-data-center-expert/. Accessed 20 Oct 2023
2. Deshmukh, A.D., Shinde, U.B.: A low cost environment monitoring system using Raspberry Pi and Arduino with Zigbee. In: International Conference on Inventive Computation Technologies (ICICT) (2016). https://doi.org/10.1109/INVENTIVE.2016.7830096
3. Krishnamurthi, K., Thapa, S., Kothari, L., Prakash, A.: Arduino based weather monitoring system. Int. J. Eng. Res. Gen. Sci. **3**(2), 452–458 (2015). http://pnrsolution.org/Datacenter/Vol3/Issue2/64.pdf. Accessed 20 Oct 2023
4. Zafar, S., Miraj, G., Baloch, R., Murtaza, D., Arshad, K.: An IoT based real-time environmental monitoring system using Arduino and cloud service. Eng. Technol. Appl. Sci. Res. **8**(4), 3238–3242 (2018). http://www.etasr.com/index.php/ETASR/article/view/2144. Accessed 20 Oct 2023
5. Vitorino de Sousa, F., Silva Araújo, R.T., Aires da Silva Alencar, J.J.: Controle das variáveis ambientais de um data center utilizando softwares e ferramentas livres. Revista da SBA, vol. 2, no. 1 (2020). https://www.sba.org.br/open_journal_systems/index.php/cba/article/view/1318. Accessed 20 Oct 2023
6. Sah, S., Majumdar, A.: Data centre temperature monitoring with ESP8266 based wireless sensor network and cloud based dashboard with real time alert system. In: IEEE Devices for Integrated Circuit (DevIC) (2017). https://ieeexplore.ieee.org/abstract/document/8073958. Accessed 20 Oct 2023

7. NodeMCU development kit v1.0 schematic. https://github.com/nodemcu/nodemcu-devkit-v1.0/blob/master/NODEMCU_DEVKIT_V1.0.PDF. Accessed 20 Oct 2023
8. IoT-Watcher reposigory. https://gitlab.citic.udc.es/a.mosteiro/tfm. Accessed 20 Oct 2023

Internet of Things Remote Laboratory for MQTT Remote Experimentation

Jesús Anhelo, Antonio Robles, and Sergio Martin(✉)

Universidad Nacional de Educación a Distancia, Madrid, Spain
smartin@ieec.uned.es

Abstract. Remote laboratories have matured substantially and have seen widespread adoption across universities globally. This paper delineates the design and implementation of a remote laboratory for Industry 4.0, specifically for Internet of Things. It employs Raspberry Pi and ESP8266 microcontrollers, to bolster online Internet of Things (IoT) learning and experimentation platforms. Such platforms hold significant value in delivering high-quality online education programs centered on IoT. Students have access to a web interface where they can write Arduino code to program the behavior of each one of the nodes of an Internet of Things scenario. This setup allows them to remotely program three NodeMCU boards in a manner akin to the usage of the Arduino IDE connected to an Arduino board locally. The system offers the ability to compile and upload code, complete with error notifications. Additionally, it furnishes several functionalities such as the ability to load new local code, save the authored code to one's personal computer, load predefined examples, access a serial monitor, and avail the Node Red platform. This amalgamation of features promises to offer a comprehensive and interactive remote learning experience for students engaging with IoT technologies.

Keywords: Remote Laboratories · On-line labs · Internet of Things · Raspberry Pi · NodeMCU · ESP8266 · Node-Red · Arduino · HTLM · Javascript · PHP · MQTT · Mosquitto Broker

1 Introduction

As the technological landscape advances at an unprecedented pace, the fourth industrial revolution, often referred to as Industry 4.0, has emerged as a transformative force driving digital innovation across global industries. Central to this movement is the Internet of Things (IoT), which interconnects a plethora of devices, creating intelligent networks and systems that are capable of perceiving, learning, and acting autonomously.

The seamless integration of physical and digital spaces offered by IoT has substantial implications for sectors ranging from manufacturing to healthcare, logistics, and beyond. Through enhanced connectivity, data sharing, and real-time analytics, IoT provides an impetus for innovative approaches to process optimization, predictive maintenance, and decision-making strategies that are set to reshape business models and market dynamics. Therefore, the understanding and application of IoT in the context of Industry 4.0 is crucial in the current era of technological advancement.

J. Bravo and G. Urzáiz (Eds.): UCAmI 2023, LNNS 841, pp. 166–174, 2023.
https://doi.org/10.1007/978-3-031-48590-9_16

However, the advent of such complex and interwoven systems demands a well-prepared workforce that can navigate the intricacies of these technological transformations. High-quality education is essential in preparing both future professionals and current practitioners to face these emerging challenges and capitalize on the opportunities they present. Particularly in light of the ongoing pandemic, online educational programs have become an imperative tool for disseminating knowledge and skills to large, geographically dispersed populations.

In this context, online laboratories play an integral role in providing hands-on, practical experience in the realm of IoT within Industry 4.0. These virtual learning environments replicate real-world systems and allow for interactive, experiential learning that goes beyond traditional, lecture-based pedagogy. They facilitate the exploration and understanding of IoT systems' design, deployment, and management, fostering critical competencies required in today's digitized industrial landscapes.

This paper delves into the intersection of Industry 4.0, IoT, and the vital role of online laboratories in providing quality education on these subjects. By illuminating the pivotal function of such platforms in driving successful digital transformation, we aim to contribute to the literature on Industry 4.0 and IoT education and underscore the necessity for robust, immersive online learning experiences in cultivating the workforce of the future.

The main aim of this paper is the description of a remote lab for Industry 4.0 based on microcontrollers programming, which allows students to remotely program an IoT scenario, learning the main concepts involved in a practical way.

2 Remote Laboratories

To address these educational challenges, the concept of remote laboratories was introduced. These are physical laboratories that can be controlled remotely, in contrast to virtual labs that merely simulate the intended environment. Remote laboratories offer several benefits [1]:

- Accessibility: Remote labs are available 24/7 from anywhere with an internet connection, offering students the flexibility to access the laboratory according to their convenience.
- Collaboration: These labs can be easily shared among different universities, with numerous projects and consortiums formed specifically for this purpose.

Despite these advantages, remote labs present a significant challenge - supervision. In the absence of physical staff to aid and guide students, the reliance on automated support systems is high. While some progress has been made in this domain, there is still a substantial need for improvement. Ideal remote laboratories should possess the following characteristics [1]:

- High Availability: The lab should be accessible without any time restrictions.
- Concurrency: The lab should have the capability to support multiple users simultaneously.
- Cost-effectiveness: The construction and operational costs of the lab should be as minimal as possible.

- Resilience: The lab must be designed to handle misuse effectively.

One prevalent approach to creating such remote labs involves the use of embedded web-based remote monitoring systems, which have several advantages over traditional PC servers:

- Cost Efficiency: This solution is cost-effective as it utilizes inexpensive hardware and open-source software.
- Low Power Consumption: The power requirements for these systems are minimal.
- Compact Size: The equipment involved is small in size.
- Security: As the web server operates using HTML and PHP on port 80, the firewall remains secure.
- Scalability: This type of remote lab can support multiple users.
- Remote Programmability: The lab can be programmed remotely, adding to its convenience.

3 Development of IoT-MQTT Remote Laboratory

In this chapter it is described the design and development of an IoT remote lab for MQTT experimentation to allow students to program and interact with the real devices over the Internet. The system is designed to address the defined challenges:

- High Availability: The remote lab Will be available 24/7 without any time restrictions, so students can practice their skills following a flexible learning approach.
- Concurrency: The lab manage multiple users thanks to a queue, so the first user to arrive have the control on the lab for some minutes.
- Cost-effectiveness: The designed software has been released as open source, the cost of the lab only depends on the hardware design.
- Resilience: The lab has been designed to handle misuse, so some actions that may hurt the hardware components are not allowed by software.

The remote lab is based on a previous version of the authors [3] and the conclusions obtained in related developments [4, 5] with the addition of new hardware, software functionalities and learning outcomes.

The architecture of this new IoT lab is shown in Fig. 1. It is based on a Raspberry Pi with Apache2 software as HTTP server connected to three nodeMCU boards based on the microcontroller ESP8266. They can be monitored through a web cam.

The remote laboratory provides a website with an Arduino code editor, where the student can write the code to be executed in each one of the 3 boards (Fig. 2); and a webcam that shows the result of such execution on the visual peripherals (LEDs, display).

The software supporting the remote lab is UNED Arduino Remote Lab, based on PHP, already described by authors in [3]. This paper presents a different experiment (focused on MQTT experimentation) with different boards (NodeMCU instead of Arduino) and electronic components. Thus, different IoT scenarios are designed. Also, for the development of this experiment, the serial monitor functionality has been implemented, to allow users communicate with the boards through serial. This is extremely useful for debugging purposes.

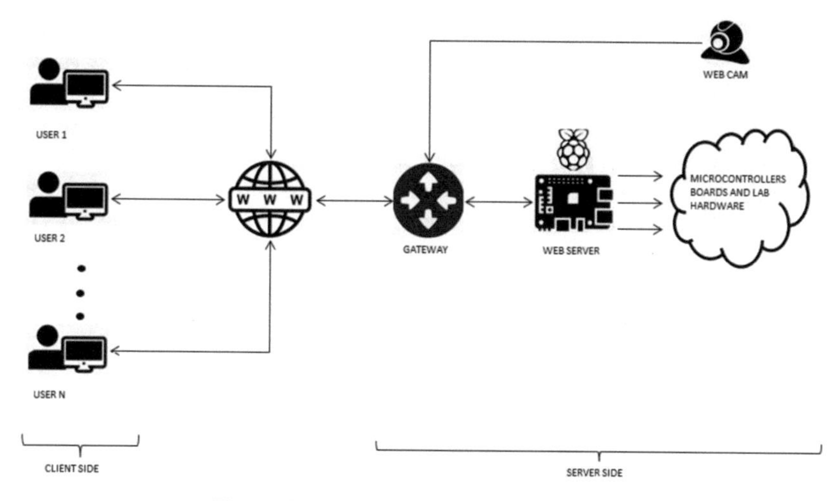

Fig. 1. Client-Server Architecture of the Lab.

Fig. 2. Screenshot of the web interface of the remote lab, including code editor (left) and webcam video stream (right).

Another improvement of this remote lab is the installation of Eclipse Mosquitto MQTT Broker. It is an open-source message broker that implements the MQTT protocol versions 5.0, 3.1.1 and 3.1. It is lightweight and is suitable for use on all devices from low power single board computers to full servers. Thanks to this MQTT broker, different MQTT practices can be carried out remotely.

The lab consists of a Raspberry Pi connected, via USB, to three breadboards with NodeMCU boards, LEDs, humidity and temperature sensors, light intensity sensors, a buzzer, and a display.

The Raspberry Pi will communicate with the NodeMCU boards through USB connections and the NodeMCUs will communicate with each other through Wi-Fi connection. As Fig. 3 shows, each one of the boards have connected several electronic components. The breadboards will be named as follow: Display Board; Sensors board 1; and Sensors Board 2.

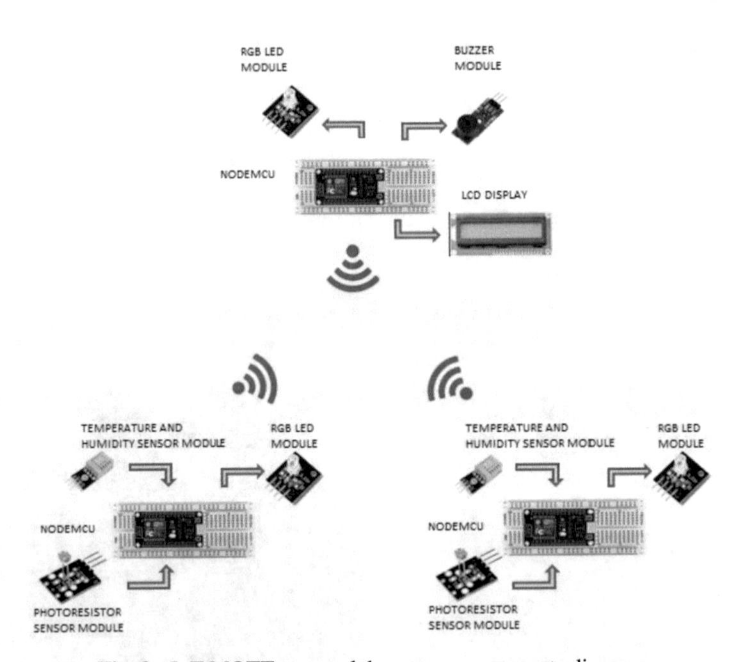

Fig. 3. IoT-MQTT remote laboratory components diagram.

4 Practices Designed for MQTT Remote Experimentation

In this section some exercises are proposed to the students so they can learn how to program an IoT environment using the MQTT protocol in a progressive way. The practices are organized in 6 different scenarios:

1. Scenario 1. Lear how to program the different electronic components of the boards with no Wi-Fi connection. Several practices are designed in this scenario:
   a. Using the LED. The aim of this first practice is to practice with PWM programming through changing a LED intensity progressively.
   b. Using buzzer. The goal of this practice is the use of the buzzer KY-006. The intensity of the module KY-016 is increased until a certain level is reached when an alarm is activated, and the buzzer sounds, and the LED is set in red color.

  c. Using LCD with I2C. In this practice a LCD1602A display with I2C bus will be used. The intensity in the led will increase until a limit in which the led will display in red and an audible alarm will sound. The display will show the instant intensity value until the alarm value from which the display will show the text "alarm".
  d. Using DHT11. The goal of this practice is the use of the humidity and temperature sensor DHT11. The output of this module is a digital signal. The led will be on in green color indicating the system is running. The data of temperature and humidity will be printed in the serial monitor.
2. Scenario 2. Master node creates Wi-Fi network. In the second scenario the ESP8266 will be used in Access Point mode (AP) to generate its own WIFI network. Several practices are designed in this scenario:
  e. One board will be the Access Point to generate de network and the rest of devices will be in Station mode and can connect to this network to communicate data. This is an access point enabled by software in a device that was not originally created to be a router and it is called "SoftAP". The access point device acts as the hub of all communications. It is the TCP/IP router to the rest of the network. Packets bound for devices within the WLAN need to go to the correct destination. The SSID keeps the packets within the correct WLAN, even when overlapping WLANs are present. However, there are usually multiple access points within each WLAN, and there must be a way to identify those access points and their associated clients. This identifier is called a basic service set identifier (BSSID) and is included in all wireless packets.
  f. Generate a Wi-Fi network with a NodeMCU in Access Point mode and two NodeMCUs in Station mode with static IP. This exercise is like the previous one except that a static IP is stablished for each board. A static IP is useful to access the device without the necessity of connect a computer to the device for knowing the IP.
3. Client-Server Communication between 2 nodes connecting to the Server's network with basic actions. In this scenario two boards will communicate through a client-server connection over HTTP in the Wi-Fi network created by the Access Point. The client will send a request to the server to the server carry out an action. The client will request through the GET method the root URI / and the server will send the "Hello" message to the client which we can see in the serial monitor and in the display.
4. Server-Client Communication between 2 nodes connecting to the Client's network with advanced actions. In this scenario two boards will communicate through a client-server connection over HTTP. A board, as server, will collect temperature, humidity and light intensity data from the environment and will send these data to the other board, as client. The client node will display the information in a LCD and will send a command to the server to turn the led on in green color in the server board during each lecture. The LED in the Client board will be turn on as well.
5. Server-Client Communication between 3 nodes connecting to the server board's network with advanced actions. In this scenario the board with the LCD will be the Server and Access Point. The other two boards will be Clients and Stations in the Wi-Fi network collecting temperature, humidity and light intensity values and sending them to the Server (Fig. 4).

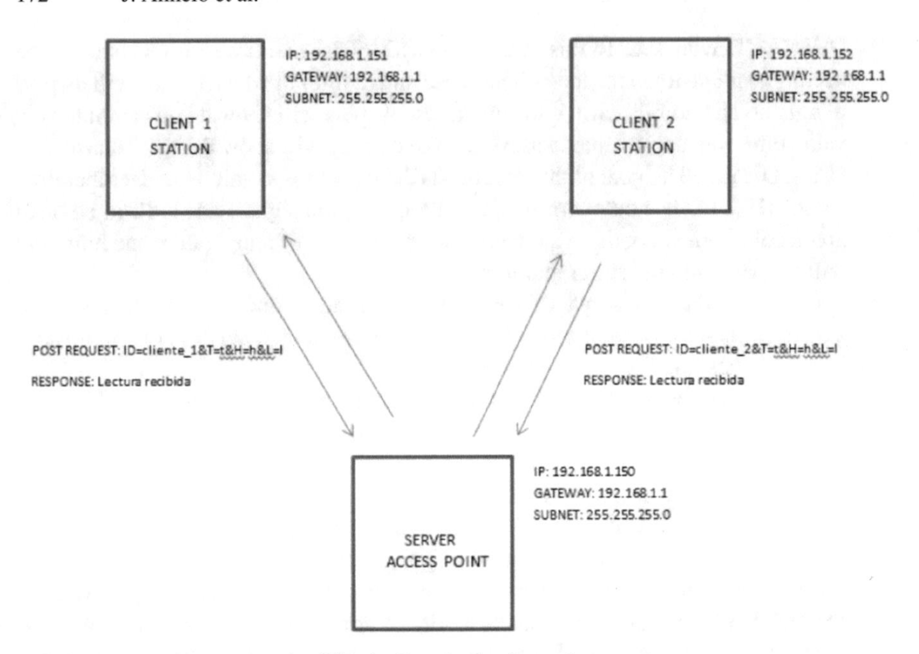

Fig. 4. Boards Configuration.

6. MQTT Communication with one NodeMCU board as broker. The task related to this scenario is the task 11. In this scenario the measurements of the sensors will be communicate via MQTT protocol with a nodeMCU board acting as broker. The Broker routes messages between clients. Each client device which connects to an MQTT broker is asked to provide a unique client identifier. The broker uses this unique client identifier to track clients and push messages to them. The main concept to understand in MQTT is "topics" – these are channels of information which clients can publish to or subscribe to messages from. Topics in MQTT are organized in a hierarchy separated by forward slashes, e.g.: iot_widgets/lightbulbs/light_88234323. An MQTT message simply consists of a topic (string) and a payload (byte array). It is up to the clients to decide what conventions to follow for payloads. Once a client device is connected to an MQTT broker, it can publish a message at any time by specifying the topic and payload. Once a client device is connected to an MQTT broker, it can also subscribe to a topic by specifying the topic and a callback function which will be run every time someone publishes a message onto the topic (Fig. 5 and 6).

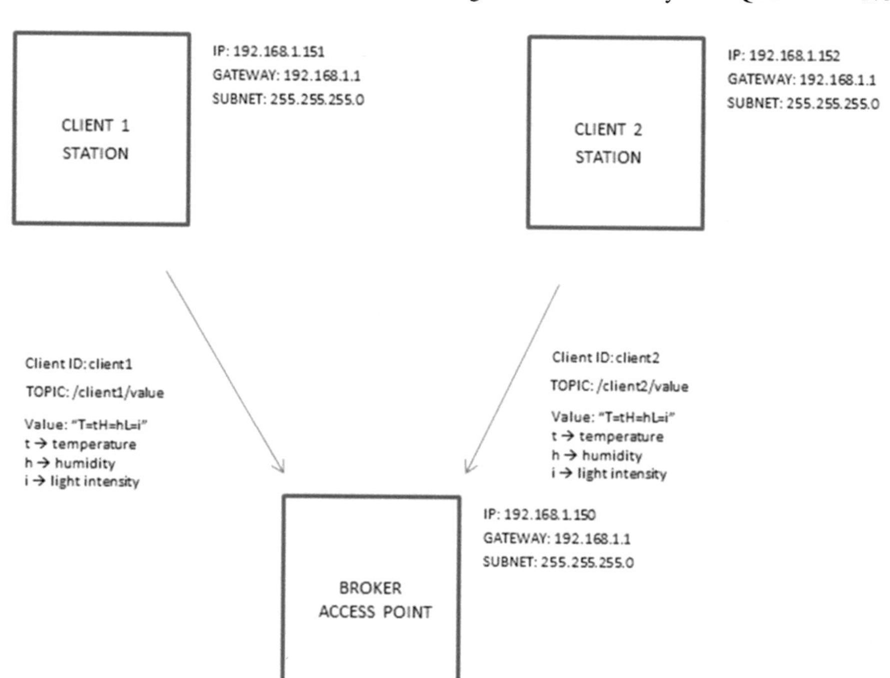

Fig. 5. MQTT subscriber/publisher architecture.

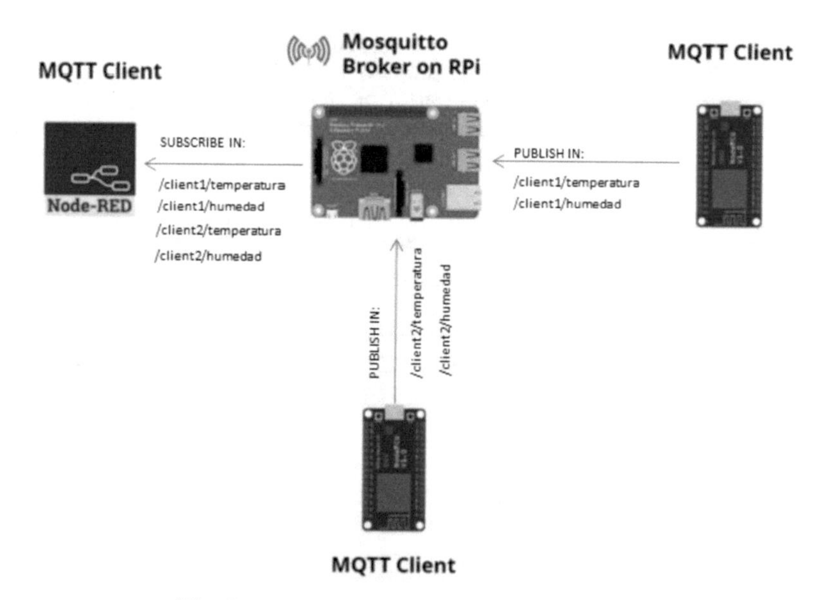

Fig. 6. MQTT operation with NodeRed as a client.

5 Conclusions

The main goal of this paper was the development of a remote IoT Laboratory based on the Arduino language, a Raspberry PI as server and boards with ESP8266 microcontrollers with the aim of supporting MQTT remote programming of real devices.

Although the objectives have been met, some limitations have been identified that will lead to new developments. For example, the need of an easier maintenance and deployment system, maybe based on virtualization techniques, such as Docker. Also, a booking system is needed if the system is expected to be used in massive environments, such as MOOCs [6]. Currently, the system solves the concurrency through a queue (first to arrive, first to get the service). However, in advance scheduling is needed for its generalization and use in massive environments.

Acknowledgments. This publication is part of the In4Labs project with reference TED2021-131535B-I00 funded by MCIN/AEI/https://doi.org/10.13039/501100011033 and by European Union "NextGenerationEU"/PRTR.

References

1. Sancristobal, E.: Virtual and remote industrial laboratory: integration in learning management systems. IEEE Ind. Electr. Mag. **8**(4), 45–58 (2014). https://doi.org/10.1109/MIE.2012.2235530
2. Blazquez-Merino, M., et al Use of VISIR remote lab in secondary school: didactic experience and outcomes. International. Conference. Remote Engineering. and Virtual Instrumentation, pp. 69–79, 2018. DOI: https://doi.org/10.1007/978-3-319-95678-7_8
3. Fernández-Pacheco, A., Martin, S. and Castro, M. Implementation of an arduino remote laboratory with raspberry pi.In: IEEE Global Engineering. Education. Conference, 2019, pp. 1415–1418. doi: https://doi.org/10.1109/EDUCON.2019.8725030
4. MARTIN, S., GOMEZ, A., CASTRO GIL, M.: Industry 4.0 remote lab based on PIC microcontrollers. DYNA **98**(4), 334–334 (2023). https://doi.org/10.6036/10925
5. MARTIN, S., GOMEZ, A., CASTRO GIL, M.: Development of an IoT remote laboratory based on PIC microcontrollers. Dyna New Technol. **10**(1), [10P.]-[10P.] (2023). https://doi.org/10.6036/NT10828
6. Martin, S. et al.,: Assessment and recognition in technical massive open on-line courses with and without on-line laboratories, In: 2023 IEEE Global Engineering Education Conference (EDUCON), Kuwait, 2023, pp. 1–4, doi: https://doi.org/10.1109/EDUCON54358.2023.10125261

Prototype of a Breathalyzer System with Biometric Filters Based on Internet of Things for a Vehicle Blocking

Moises Rodriguez-Cruz, Miguel A. Wister(✉), Ernesto Rafael Leon-Cornelio, and Jose A. Hernandez-Nolasco

Juarez Autonomous University of Tabasco, Carr. Cunduacan-Jalpa Km. 0.5, 86690 Cunduacan, Tabasco, Mexico
{222H21003,231H18002}@alumno.ujat.mx, mwister@hotmail.com, adan.hernandez@ujat.mx

Abstract. Some devices built into automobiles assess the driver's alcohol status to deny or allow engine start; however, these devices do not identify the person taking the breathalyzer test, making it easy for someone else to perform a fraudulent test. This article proposes an architecture based on IoT that integrates a biometric sensor, an alcohol sensor, and a GPS module; connected to a microcontroller capable of implementing the MQTT protocol to transmit the data obtained to a server in case the allowed limit of 0.4 mg/lt is exceeded, sends an alert via WhatsApp to an emergency number with the location of the vehicle. In the experimental design, ten breathalyzer tests were carried out on drunk people whose results showed the validation of the driver's identity through a biometric sensor. In addition, this project aims to implement security filters through biometric sensors to prevent identity theft by another person.

Keywords: Internet of Things · Raspberry Pi 4 · MQTT · vehicle blocking · alcohol

1 Introduction

Automobiles have become a fundamental part of daily life. Human beings need to move from one place to another. However, they can cause other types of risks that are not inherent to vehicles but rather the misuse that they are given. These problems are car accidents, which bring monetary losses due to collisions with property and have even caused the death of people, drivers, passengers, and bystanders. These types of accidents are caused by various factors, such as driving tired, distracted drivers, and using a vehicle if a driver has consumed illegal substances such as drugs or alcohol. Devices called breathalyzers detect the levels of alcohol through a driver's exhaled air to determine if the user is

M. Rodriguez-Cruz, M. A. Wister, E. R. Leon-Cornelio and J. A. Hernandez-Nolasco—These authors contributed equally to this work.

J. Bravo and G. Urzáiz (Eds.): UCAmI 2023, LNNS 841, pp. 175–180, 2023.
https://doi.org/10.1007/978-3-031-48590-9_17

intoxicated. [1] shows the levels of alcohol allowed for the driver in different countries. Some systems, such as Alcolock [2], perform breathalyzer tests to prevent a vehicle from starting if a person is within the permitted alcohol limits. Therefore, developing this proposed architecture is feasible since these devices [2] exist. This research contributes to including new variables that act as filters to determine if the driver performs the test and cannot perform identity theft.

2 Related Works

One of the oldest uses of healthcare technology has been the development of blood alcohol content tests to detect impaired driving [3]. Over time, devices called breathalyzers were created where breathalyzer tests are performed on the breath exhaled by a driver. Despite this, new methods have been contemplated where IoT systems are used to prevent drunk driving [4]. The implementation of sensors for the activation/deactivation of the ignition system and geographic location of a vehicle using integrated GPS coordinates through a local WiFi network [5]. Biometric identification, especially through fingerprint scanning, is used to authenticate users before allowing them to start the vehicle. It is verified if the fingerprint matches those authorized in a database. If there is a match, the ignition of the vehicle is allowed. Otherwise, the user must scan a valid fingerprint to start the vehicle [6]. [7] It proposes a system that has the ability to register multiple users using an Atmega 328 microcontroller and an esp8266 WiFi module. The fingerprint detector is connected to the microcontroller, along with an LCD screen, buttons, and a starter motor. Ten unique fingerprints were enrolled and programmed into the system, with ten attempts each, and 82 successful valid fingerprint scans were obtained out of a total of 100 attempts. The errors in the scans were due to the different ways the finger was placed on the scanner. [8] Together with a fingerprint scanner and an MQ3 alcohol sensor, a network of sensors connected to the Arduino board will be made. This board sends fingerprint and alcohol test results to the Raspberry Pi via a serial connection.

3 Design

This section details the materials and equipment used in the design and general technical aspects. The architectural proposal comprises four sensor nodes: fingerprint sensor, MQ3 sensor, GPS sensor, and strain gauges sensor, each of which monitors variables independently. The sensors are connected to the ESP8266 board in a wired manner, using the corresponding pins according to the power configuration and the pins declared in the program for data acquisition. As for the communication between these nodes and the Raspberry Pi, this is done wirelessly through the MQTT protocol since the Raspberry Pi has been configured as an MQTT broker. Once the data is received on the computer (Raspberry Pi), it is evaluated in the same program to determine if it should activate the engine or, failing that, deny ignition. In addition, a message is sent to an emergency telephone number with the vehicle's location if necessary.

3.1 Materials and Equipments

Table 1 shows all materials used for the development of this architecture.

Table 1. Architecture materials

Item	Description	Model/Type	Qty.
1	ESP8266	NodeMCU	3
2	Raspberry Pi board	V4 Model B	1
3	SD-card	32 Gb	1
4	MQ Sensor	MQ-3	1
5	Smartphone	Samsung S22 Ultra	1
6	GPS Module	Neo 6m	1
7	Fingerprint Sensor	As608	1
8	OLED Display Screen	128 × 64 0.96 for Arduino	2
9	Strain Gauges	50Kg each	4

The general architecture consists of four primary nodes: MQ-3 sensor node, GPS node, fingerprint sensor node, and strain gauges. These nodes are represented in Fig 1, which are essentially microcontrollers connected to the ESP8266 modules. Its function is to measure various variables, the MQ-3 node measures the Blood Alcohol Level (BAC) through exhaled air, the GPS node sends the location of the vehicle through coordinates to a WhatsApp number, and the fingerprint node captures a detailed image of the footprint, recording the grooves, ridges and characteristic points. These data are transmitted via the MQTT protocol and received by a central computer acting as an MQTT broker. Also, the data sent by these nodes is stored within the said main computer, using a specific database. In this case, SQLite is used. In parallel, depending on the registered BAC level, it is decided whether or not a WhatsApp message will be sent to the emergency contact, including the vehicle's location.

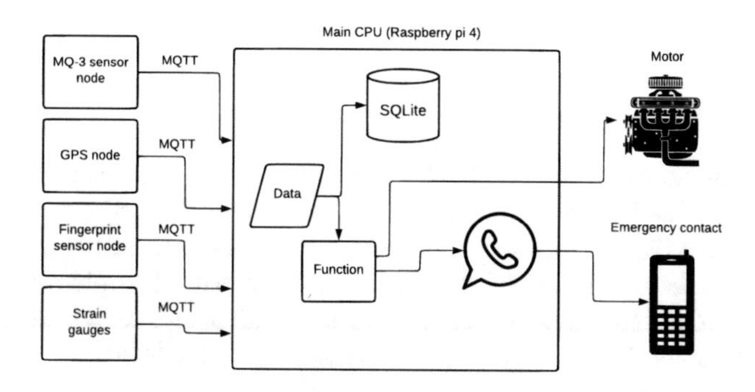

Fig. 1. Conceptual Architecture

4 Testing and Implementations

In this phase of the study, the system is partially implemented in order to verify the operation of the MQ-3 sensor node, the GPS node, and the fingerprint sensor node in addition to MQTT communication and the WhatsApp Bot.

As previously mentioned, it is required to obtain the alcohol level present in the driver; for this, the MQ-3 (Fig. 2a) sensor was used in conjunction with an esp8266 to calibrate the sensor measurement the "MQUnifiedsensor.h" library was used. This library makes it easier For the calibration, and we only have to pass the data obtained by the sensor with clean air; the alcohol data is obtained through an exponential regression that is established in the same library.

The GPS node (Fig. 2b) is responsible for establishing communication with a GPS module and performing the analysis of the NMEA frames received. The latitude and longitude coordinates representing the geographic position are extracted from these frames.

The fingerprint node (Fig. 2c) performs two functions (fingerprint registration and fingerprint validation) in two phases. For them, the Adafruit_Fingerprint library stores fingerprints on the device and performs real-time matching to verify the identity of users. This enables fingerprint verification and registration functionality in the system.

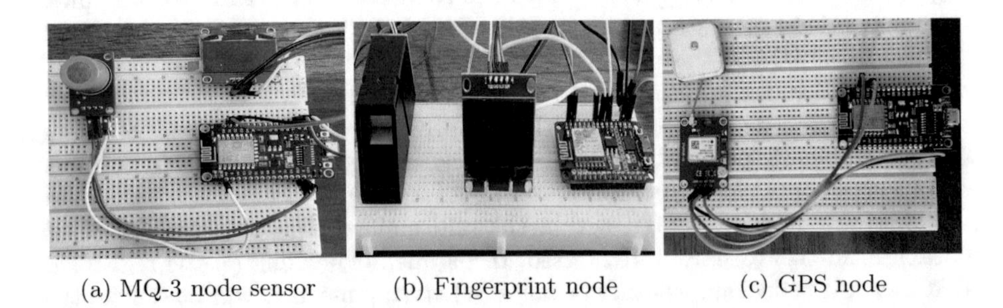

(a) MQ-3 node sensor (b) Fingerprint node (c) GPS node

Fig. 2. Sensor circuits used in the prototype

5 Results

To evaluate the nodes of the architecture, a different test was carried out on each one: With the MQ-3 sensor node: (Fig. 3a), tests were carried out to determine the measurement; these tests were carried out with ten test subjects, from whom a sample was taken before drinking alcohol and another sample after ingesting alcohol, to validate the data obtained. Through the MQ-3 sensor node, another sample was taken with a commercial-use breathalyzer (At6000); the results show that there is a margin of error of 1 to 2 g/210L.

The fingerprint sensor node (Fig. 3b) was evaluated in two cases: registration and identity validation. The index finger was placed vertically for registration to capture and save the fingerprint in the sensor's memory. A name and an ID corresponding to a user were registered. In validation tests, the sensor correctly identified the enrolled finger. In summary, the sensor works well when recognizing the registered fingerprint, but it shows limitations when identifying different fingerprints from people not registered or in positions other than the established one.

The GPS node (Fig. 3c) was tested using the WhatsApp Bot as an interface. Five locations were evaluated to verify the accuracy and the time to obtain the coordinates. The results showed that obtaining coordinates varied slightly depending on the location, with a maximum time of 2 min in the most extended case. Once activated, the GPS node worked correctly and provided coordinates accurately.

The message generated by the WhatsApp Bot includes a link like the following shown: "http://maps.google.com/maps&z=15&mrt=yp&t=k&q=18.075560,-93.170570" which directs Google Maps with the coordinates of the node, which allows easy visualization of the location. The WhatsApp Bot sent the location without any details. In short, the GPS node demonstrated reliable performance in obtaining coordinates, and the interaction with the WhatsApp Bot effectively facilitated access to location information.

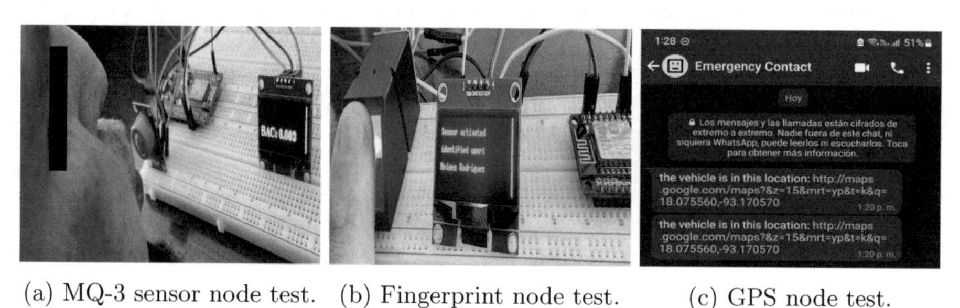

(a) MQ-3 sensor node test. (b) Fingerprint node test. (c) GPS node test.

Fig. 3. Output message of the proposed systems

6 Conclusion

So far, tests have been carried out with the previously mentioned nodes. Significantly, the MQ-3 sensor node can improve reading accuracy. In addition, the possibility of repeating the tests using a more sophisticated and validated breathalyzer is contemplated. It is worth mentioning that, in the current difficulties, a conventional breathalyzer was used.

On the other hand, the nodes used have demonstrated effective performance in fulfilling their operation. As the next advance, a scale will be incorporated to provide additional data about the driver and evaluate whether it complements the driver's identification. Once all the parts of the architecture are integrated, a database will be implemented to collect the data obtained by each previously mentioned node.

This architecture will have the capacity to expand and add more nodes that contribute to the identification of the user and to verify if they are in optimal conditions to drive the vehicle that incorporates this system. In addition, this approach will also contribute to improving automated driving systems.

References

1. COPRO: ley de Conducir Borracho Por País. https://copro.com.ar/Ley_de_conducir_borracho_por_pais.html
2. autocasion: Alcolock. https://www.autocasion.com/diccionario/alcolock
3. Sarkar, T., Shaw, S.: IoT based intelligent alcohol detection system for vehicles. In: Proceedings of the 4th International Conference on Big Data and Internet of Things, pp. 1–5 (2019)
4. Farooq, J.S., Soundarya, V., Rao, V.S., Chandraprabha, K., et al.: Safe drive: an automatic engine locking system to prevent drunken driving. In: 2018 3rd IEEE International Conference on Recent Trends in Electronics, Information & Communication Technology (RTEICT), pp. 1957–1961. IEEE (2018)
5. Brito, G., Salazar, F.W., Lema, E.O., Sánchez, A.P., Pérez, H.V., Buele, J.: Vehicle locking system using an electronic breathalyzer and notification by mobile communication. J. Comput. Theor. Nanosci. **17**(1), 206–215 (2020)
6. Ofoegbu, E.O.: An adaptive user authentication architecture for drunk driving and vehicle theft mitigation. Int. J. Eng. Manuf. **12**(6), 32 (2022)
7. SriAnusha, K., Saddamhussain, S., Kumar, K.P.: Biometric car security and monitoring system using IoT. In: 2019 International Conference on Vision Towards Emerging Trends in Communication and Networking (ViTECoN), pp. 1–7. IEEE (2019)
8. Ramamurthy, B., Latha, N.A.: Development and implementation using Arduino and raspberry Pi based ignition control system. Adv. Comput. Sci. Technol. **10**(7), 1989–2004 (2017)

Energy Aware Systems, Communications and Security

Bio-Inspired Security Analysis: A Domotic Use-Case

Vincenzo Conti[1](✉), Alberto Gallo[2], Mauro Migliardi[3], and Salvatore Vitabile[2]

[1] Universitá degli Studi di Enna Kore, Enna, Italy
vincenzo.conti@unikore.it
[2] Universitá degli Studi di Palermo, Palermo, Italy
[3] Universitá degli Studi di Padova, Padova, Italy

Abstract. In a world that is heavily relying on connected computers for the efficient execution of most daily tasks, Computer Security is absolutely critical. Therefore, in order to perform a complete analysis, new models and paradigms are needed to better manage the complexity of systems for an automated and data-driven economy. In past work we have described a bio-inspired approach that leverages metabolic networks to enhance and facilitate the use of attack-graph analysis to evaluate the security of systems, namely the BIAM framework. In this paper we describe the application of the BIAM framework to the search, analysis and assessment of the vulnerabilities of a simulated real-world use-case in the field of home-automation and ambient-intelligence.

Keywords: Attack Graphs · Systems Security · Network Security · Bio-inspired Techniques · Ambient Intelligence

1 Introduction

Today, with an infrastructure that evolves towards the use of sensors, data services to be managed remotely, and systems that interact with each other, a main problem arises relating to information security. In fact, the growing complexity of the systems makes them increasingly similar to interconnected networks and networks of networks; this fact lead us to study and analyze security by exploiting well-known approaches and techniques already used for such complex networks. A promising approach for the analysis of complex networks is attack graph analysis, where attack graphs are special graphs that mathematically describe attack scenarios [1]. An attack graph is the sequence of events and actions needed to carry out a cyber attack. There are different types of attack graphs, each of which represents a particular information about an attack. In various models, nodes can represent the target of the attacker or the phases of an attack. In general, nodes are associated to a cost, a cause or a probability of success. However, the manual analysis of attack graphs is tedious and error prone, hence, in this work we use a paradigm that focuses on representing computer/sensors network as a

J. Bravo and G. Urzáiz (Eds.): UCAmI 2023, LNNS 841, pp. 183–194, 2023.
https://doi.org/10.1007/978-3-031-48590-9_18

biological network where security characteristics are used to model the "security metabolism" of the system. With this representation, it is possible to study the system using bio-inspired algorithms and to analyze and evaluate its robustness from a security point of view [2]. Furthermore, the application of this paradigm allows identifying critical paths that present the highest level of vulnerability for the system.

In this paper, after describing the main characteristics of a software framework dedicated at performing bio-inspired analysis of complex networks, namely the BIAM framework [3], we will perform the analysis and the assessment of the vulnerabilities in an almost real-world use-case in the field of home-automation and ambient-intelligence.

In order to apply the metabolic network model to attack graphs analysis, we leverage four interacting bio-inspired algorithms: topological analysis, extreme pathway analysis, Dijkstra algorithm, and flux balance analysis. In literature, authors have just used these techniques in many research field, IoT scenarios [4], service-oriented networks [5], urban vehicles routing optimization [6], all types of systems that can be represented by a complex network and thus can be analyzed by bio-inspired techniques.

In this experiment we will characterize the domestic infrastructure as a network, and we will inject all the information about vulnerabilities that are available in the Common Vulnerabilities and Exposures (CVE) database. Starting from this information an attack graph will be automatically created and then it will be analyzed using a tool that allows us to obtain values that determine which is the greatest risk and the minimum number of changes needed to morph the network into a new version that could be considered at least partially safe.

The rest of the manuscript is organized as follows. Section 2 introduce to the ambient-intelligence and security; Sect. 3 briefly describe the BIAM framework; Sect. 4 describe the home automation scenarios and the performed tests; finally, the conclusions are reported.

2 Ambient-Intelligence and Security

Ambient Intelligence is the application into everyday environments to seamlessly provide assistive and predictive support in a multitude of scenarios. These can be as diverse as autonomous vehicles, smart homes, industrial settings, and so on [7]. Closely linked to these scenarios and their possible representation through complex networks is the issue of security and cyber attacks. Cyber attacks and related countermeasures have been the subject of scientific research for years, and in the last period attack graphs are a much investigated topic. Since the first proposal of the attack graph theory, done by Phillips and Swiler [8], for analyzing network vulnerabilities, several studies have been improving this research field, spacing out through a great variety of methods and issues. In literature there are several techniques used for attack graph analysis.

As mentioned in [9], an attack graph can be considered as a game to which the Stacklberg equilibrium is applied to search for the best strategy that defenders must apply to defend against attackers. As previously said by Amman et al. [10], "an attack graph quickly becomes unmanageable large as network complexity grows" because it will exponentially expand itself. Much research work tried to optimize the network graph: Zhang et al. [11] focused on how to check out some preconditions to reach a destination; Xiao et al. [12] proposed a component-centric attack graph that takes in account the preconditions of the exploits; Liu et al. [13] proposed a method that combines greedy policy, forward exploration and backward searching to reduce the complexity of the attack graph; Chao et al. [14] proposed an algorithm to generate the minimum initial set, specifically a modified heuristic algorithm based on the ant colony optimization. All the work cited above approached the dimensionality problem trying to optimize a graph by reducing its number of nodes. Another approach to improving the graph analysis, is based on using the Common Vulnerability Scoring System (CVSS) metric in order to add damage and exploitability information to the graph [15]. This last approach allows determining the damage induced by attacker to the targeted network. This approach has led to a different line of research inside which we may cite studies like the one of Malzahn et al. [16], that uses CVEs - obtained from scanning the network - to auto-check and validate the graph itself throughout an auto-exploit, or the work of Musa et al. [17] that uses CVSS base score to identify and study only the most dangerous paths. The proposed manuscript aims at offering a more complete and flexible analysis of a graph using a biological inspired viewpoint based on how chemical compounds are involved in cellular reactions in a metabolic network [2]. It allows drawing, managing, analyzing and saving a complex graph that may or may not also use CVSS metrics.

3 The BIAM Framework

The framework used in this work, namely BIAM, has been implemented and proposed in the first time in [3]. This framework allows building and handling random and small-world networks. The paradigm used is that any real computer/sensors network can be represented with a biological network and, therefore, studied using bio-inspired algorithms. For this reason, BIAM is based on four bio-inspired algorithms for networks security analysis, namely: Topological Analysis, Extreme Pathway Analysis, Dijkstra Analysis and Flux Balance Analysis. Furthermore, it is equipped with a user-friendly GUI capable of showing in real time: the generated graph, the changes made and the results obtained from the analyses. It can also scan the current net and obtain many information about components and vulnerabilities as it can.

3.1 Topological Analysis

This analysis gives a set of network structural information on its size and the information on each node. With more details, it provides the following list of parameters and coefficients: metabolites number, reactions number, network diameter, input average degree (related to the input link number of a node), output average degree (related to the output link number of a node), within-module degree, and participation coefficient [18].

3.2 Extreme Pathway Analysis

Extreme pathways are the unique set of vectors that completely characterize the steady-state capabilities of a biological network [19,20]. The steady-state operation of the biological network is constrained to the region within a cone, defined as the feasible set. In some special cases, under certain constraints, this feasible set collapse in a single point inside the cone. BIAM detects the network extreme pathways in three steps: first, user must select the directions of fluxes (incoming, outcoming or bidirectional); after the framework implements the ExPa algorithm for detecting the extreme rays/generating vectors of convex polyhedral cones; finally, the framework returns the list of extreme pathways as a function of linear combination of internal and external fluxes. Moreover, it shows the maximum length for each extreme pathway.

3.3 Dijkstra Analysis

This analysis is based on the homonymous algorithm that find the shortest path from a given source to a given destination [21]. The implementation of this function follows the pattern of a Depth First Search, so it explores each branch of the graph until it finds a path that link source and destination. It shows the maximum length of the returned path as well as the previous analysis.

3.4 Flux Balance Analysis

The Flux Balance Analysis is a mathematical technique based on the fundamental physical and chemical laws that describes quantitatively the microbial metabolisms. Cellular networks robustness can be investigated in terms of the functional state of the nodes. FBA is a constraint-based modeling approach [22]: it assumes that an organism reaches a steady-state that satisfies the physical-chemical constraints and uses the mass and energy balance to describe the potential behavior of metabolism. The space of all feasible solutions lies within a three-dimensional convex polyhedron, in which each point of this space satisfies the constraints of the system. When the system has an optimal and limited solution, this is unique and it is located on a polyhedron vertex. However, the system can have multiple optimal solutions used to detect network redundancies [22].

4 Home Automation Scenarios and Tests

The analysed network (Fig. 1 and Fig. 2) represents a home-like scenario based on a topology that diverges from the most commonly used star topology. The network consists of 3 smartphones, 1 laptop and 1 led based lighting automation gateway connected to a router which is connected to another one. The latter router is using a specific firmware and it is linked to 1 smart TV and 1 NAS. In our simulation the attacker accesses the network through the laptop (device number 4 in Fig. 1) and his target is the NAS station (device number 9 in Fig. 1). In order to generate the graph, the first step is to select all the components. In order to make the scenario viable for our analysis, we have to check that each component is present in the NIST database, and that it has some vulnerabilities. Each component presents a certain number of reported vulnerabilities (Table 1) that are characterized by a set of impact metrics. These metrics determine the severity of the vulnerability related to a certain property (confidentiality, availability, integrity).

The second step is to decide how to lead the attack through all chosen components (Table 1) and their vulnerabilities. In order to carry it out, we have to pick up the most appropriate property to crack.

This way vulnerabilities are handled based on the high severity of the chosen property. This action will alter the number of total vulnerabilities considered for each component.

Unfortunately, not all components have at least one vulnerability that fits this filtering condition, so we had to manage this issue by choosing another attack to lead the way off. Then, a new vulnerability-related graph (Table 2, Fig. 3) has been created with a number of nodes equivalent to the sum of all the CVE entries found, related to a chosen attack type and to other linked vulnerabilities (Table 3), plus a node (node number 1 in Fig. 3) which represent the intruder and a node (node number 22 in Fig. 3) which represent the final target of the attack; the weight of each edge will be determined by the normalization of the maximum CVSS score of all CVEs connected to the next Common Platform Enumeration (CPE) obtained from the database.

As happened for the components previously, each vulnerability has been reported by the online database.

Table 1. Component-Vulnerability list of the devices of the scenario represented in Fig. 1

Id	Component Name	Requirement	Num. Vulnerabilities
1	Samsung Galaxy S22	Android 12	1
2	Xiaomi Redmi Note 11	Android 11	1
3	OnePlus 3T	Android 7	7
4	Microsoft Windows11 x64	Windows 11	229
5	Ikea Tradfri Gateway E1526	Led Gateway	1
6	TP-Link ARCHER AX50	Router	1
7	Drupal 9.3.10	Firmware	197
8	TP-Link TL-SG108E 3.0	Router	2
9	Synology DSM 7.1	NAS	9
10	Sony Bravia	Smart TV	2

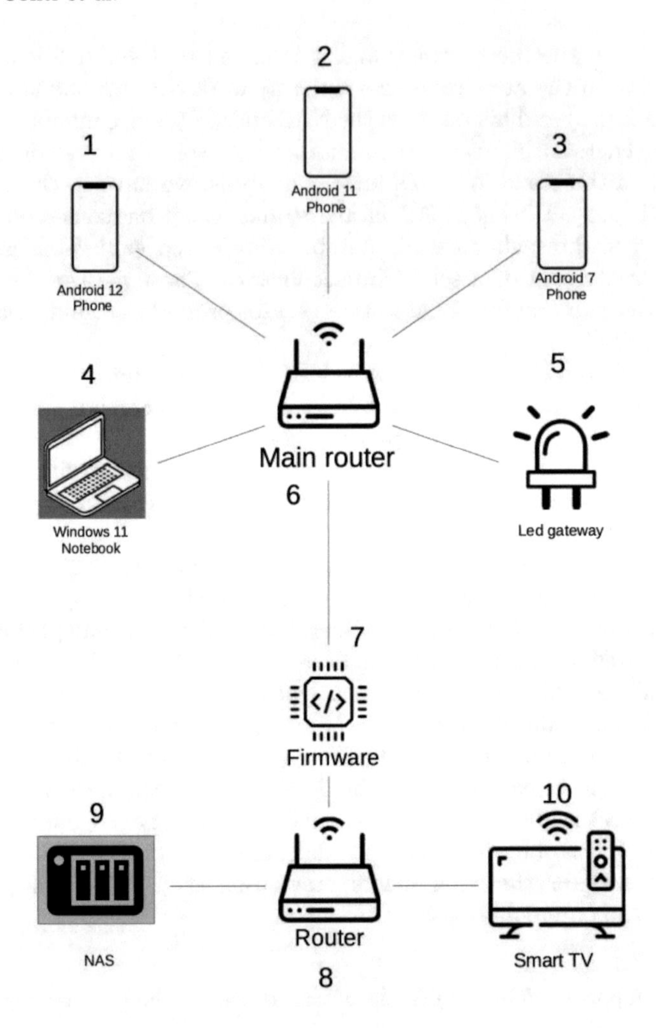

Fig. 1. Network graph

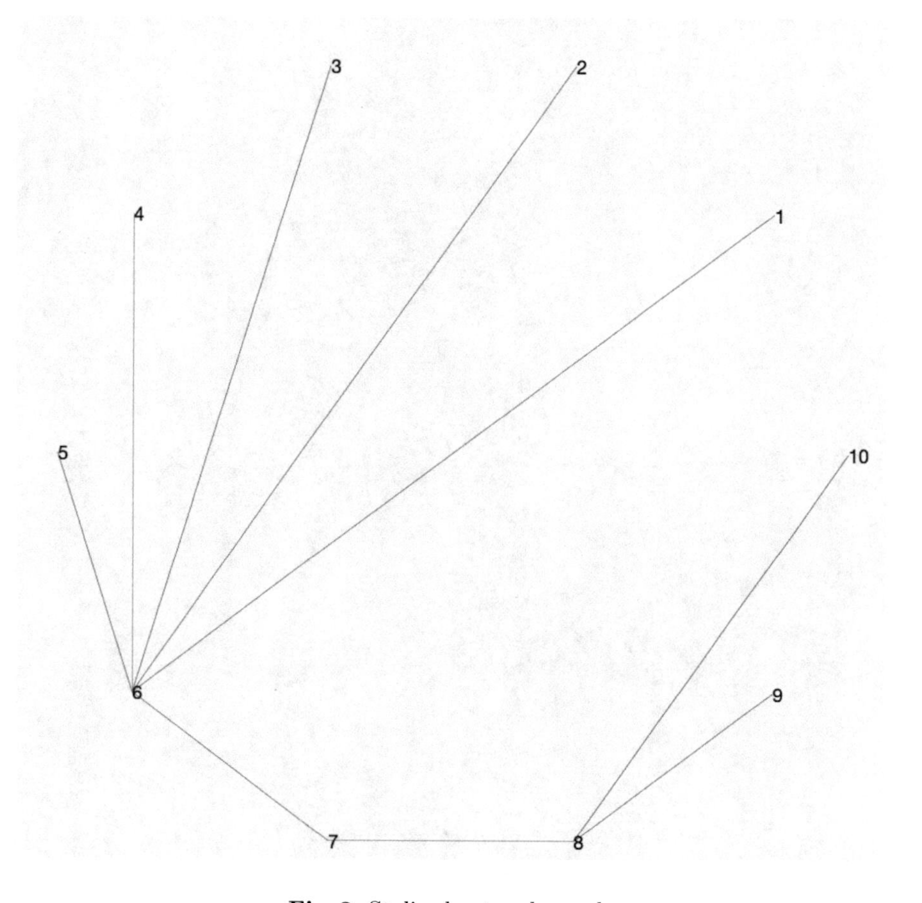

Fig. 2. Stylized network graph

Table 2. Component to vulnerability node.

Component Node	Vulnerability Node
1	2, 3
2	4, 5, 6
3	7, 8
4	9, 10, 11
5	12
6	13, 14
7	15, 16
8	17, 18
9	19
10	20, 21

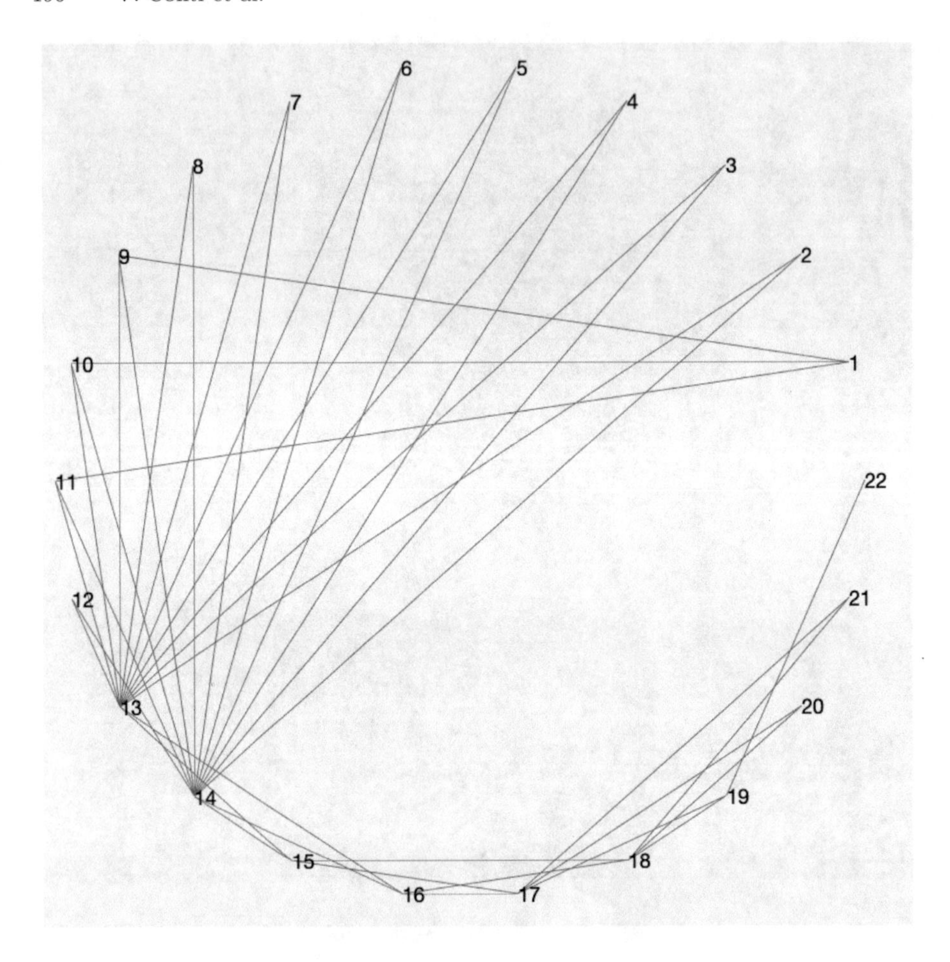

Fig. 3. Stylized vulnerability-related graph.

Table 3. Component-Vulnerability Attack oriented list.

Id	Component Name	Deployed Attack	Final Vulnerabilities
1	Android 12	Integrity	2
2	Android 11	Availability	3
3	Android 7	Confidentiality	2
4	Windows 11	Confidentiality	3
5	Led Gateway	Availability	1
6	Router	Integrity	2
7	Firmware	Integrity	2
8	Router	Integrity	2
9	NAS	Confidentiality	1
10	Smart TV	Availability	2

Now the graph is ready to be analysed. The proposed analyses are designed to analyse metabolic network, particular networks that describe the reactions that, in the case of cellular metabolism, leads to the production of nutrients for the cell.

During the working phase of BIAM (Fig. 4) the sequence of steps is the following.

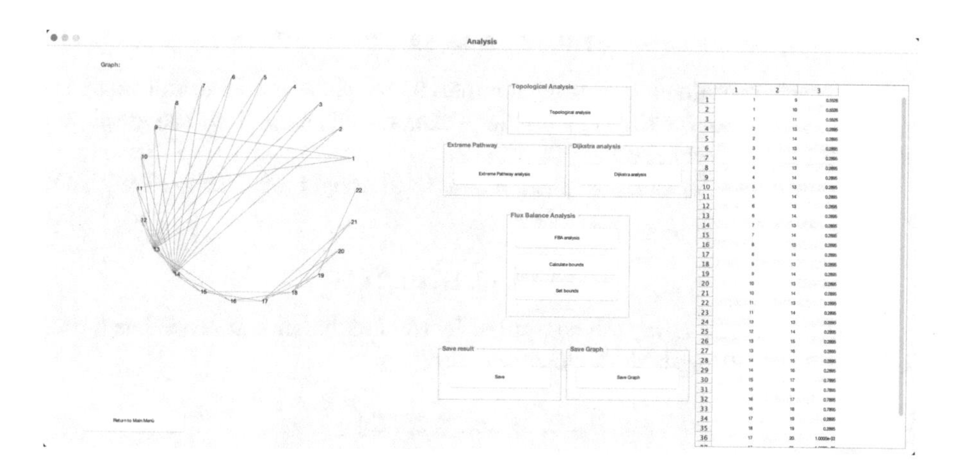

Fig. 4. BIAM Gui in working phase.

Firstly, a Dijkstra analysis is performed. This analysis use Dijkstra's algorithm to obtain the first minimum path from a source to a destination.

Then, the Extreme Pathway Analysis (ExPA) is performed. This analysis works on the stoichiometric matrix and it explores all the possible paths from the source to the destination of the attack.

Finally, the Flux Balance Analysis (FBA) is performed. This analysis calculates the optimal path and the related optimal fluxes through what we obtained from ExPA function. The result path of FBA function must have a minimum input flux (called nutrient uptake) in order to cellular survival.

Focusing on the component-related graph (Fig. 2), it is possible figure out which are the possible threats and how to possible deal with them: ideally, the most important files reside on the NAS station (device number 9 in Fig. 1), in fact it is connected to a more secure side of the network which is protected by the Drupal firmware (device number 7 in Fig. 1). So, the most threatened sub-network is the one that relies on the wi-fi router (device number 6 in Fig. 1) where all devices are connected. Based on the overall of component reported vulnerabilities, it is possible to say that the most dangerous device is the computer with Windows O.S. (device number 4 in Fig. 1).

With all those considerations, let's find out which is the most vulnerable path and how to lessen its threat.

The Dijkstra analysis returned path has a total weight of 2.53*e-10 with the following nodes:

$$[1, 9, 13, 16, 17, 19, 22]$$

Whereas best path computed by the flux balance analysis has a total weight of 2.49*e-10 with the following nodes:

$$[1, 9, 13, 15, 18, 19, 22]$$

Each path contained the node number 9, which is a vulnerability of the Windows device. So, for the next test let's exclude this node from the graph and see how results will change.

The new Dijkstra analysis path has a total weight of 5.23*e-10 with the following nodes:

$$[1, 10, 13, 16, 17, 19, 22]$$

Whereas the best new path computed by the flux balance analysis has a total weight of 5.14*e-10 with the following nodes:

$$[1, 10, 13, 15, 18, 19, 22]$$

In conclusion, the application of the proposed bio-inspired technique allowed to analyse a network, simulate an attack and figure out a way to contrast it.

5 Conclusions

In this manuscript a case study on a home automation network has been analyzed and studied adopting a bio-inspired paradigm and a software framework dedicated to apply such paradigm to the analysis of complex networks. The application of the bio-inspired paradigm and of four related analysis algorithms (namely topological analysis, extreme pathway analysis, Dijkstra algorithm, and flux balance analysis) allowed identifying the network weak points and the overall level of security using an attack graph that is a special graph describing attack scenarios. For this type of study the BIAM framework, a known framework of literature, has been used. This approach, already used in the literature in other research fields, has given satisfactory results in terms of vulnerability identification and possible interventions to increase the security level of the entire network. In future work, we plan to further automate all the steps of the study, in order to make the framework easier to use; in fact, in this version the acquisition and recognition steps, of the local network, are still not totally automatic processes and contain some manual tasks.

References

1. Alhomidi, M.A., Reed, M.J.: Attack graphs representations. In: 4th Computer Science and Electronic Engineering Conference (2012)
2. Vitabile, S., Conti, V., Lanza, B., Cusumano, D., Sorbello, F.: Metabolic Networks Robustness: theory, simulations and results. J. Interconnect. Netw. **12**(3), 221–240 (2011)
3. Conti, V., Ruffo, S.S., Vitabile, S., Barolli, L.: BIAM: a new bio-inspired analysis methodology for digital ecosystems based on a scale-free architecture. Soft. Comput. **23**(4), 1133–1150 (2019)
4. Conti, V., Ziggiotto, A., Migliardi, M., Vitabile, S.: Bio-inspired security analysis for IoT scenarios. Int. J. Embedded Syst. **13**(2), 221–235 (2020)
5. Conti, V., Militello, C., Rundo, L., Vitabile, S.: A novel bio-inspired approach for high-performance management in service-oriented networks. IEEE Trans. Emerg. Top. Comput. **9**(4), 1709–1722 (2021)
6. Vitello, G., Alongi, A., Conti, V., Vitabile, S.: A bio-inspired cognitive agent for autonomous urban vehicles routing optimization. IEEE Trans. Cogn. Dev. Syst. **9**(1), 5–15 (2017)
7. Ko, H., Ramos, C.: A survey of context classfication for intelligent systems research for ambient intelligence. Int. Conf. Complex, Intell. Softw. Intensive Syst. (2010)
8. Phillips, C., Swiler, L.P.: A graph-based system for network vulnerability analysis. In: Proceedings of New Security Paradigms, pp. 71–79 (1998)
9. Chokshi, I., Das, S., Ghosh, N., Ghosh, S., Kaushik, A., Sarkar, M.: NetSecuritas: an integrated attack graph-based security assessment tool for enterprise network. In: Proceedings of the International Conference on Distributed Computing and Networking (2015)
10. Ammann, P., Pamula, J., Ritchey, R., Street, J.: A host-based approach to network attack chaining analysis. In: 21st Annual Computer Security Applications Conference, pp. 10–84 (2005)
11. Zhang, T., Ming-Zeng, H., Li, D., Sun, L.: An effective method to generate attack graph. Int. Conf. Mach. Learn. Cybern. **7**, 3926–3931 (2005)
12. Xiao, X., Zhang, T., Wang, H., Zhang, G.: A component-centric access graph based approach to network attack analysis. In: International Seminar on Future Information Technology and Management Engineering, pp. 171–176 (2008)
13. Liu, Z., Li, S., He, J., Xie, D., Deng, Z.: Complex network security analysis based on attack graph model. In: Second International Conference on Instrumentation, Measurement, Computer, Communication and Control, pp. 183–186 (2012)
14. Chao, Z., Huiqiang, W., Fangfang, G., Mo, Z., Yushu, Z.: A heuristic method of attack graph analysis for network security hardening. In: International Conference on Cyber-Enabled Distributed Computing and Knowledge Discovery, pp. 43-47 (2014)
15. Gallon, L., Bascou, J.: CVSS attack graphs. In: Seventh International Conference on Signal Image Technology and Internet-Based Systems, pp. 24–31 (2011ğ)
16. Malzahn, D., Birnbaum, Z., Wright-Hamor, C.: Automated vulnerability testing via executable attack graphs. In: International Conference on Cyber Security and Protection of Digital Services, pp. 1–10 (2020)
17. Musa, T., et al.: Analysis of complex networks for security issues using attack graph. In: International Conference on Computer Communication and Informatics, pp. 1–6 (2019)

18. Guimerá, R., Amaral, L.A.N.: Functional cartography of complex metabolic network. Nature **433**, 895–900 (2005)
19. Schilling, C.H., Letscher, D., Palsson, B.: Theory for the systemic definition of metabolic pathways and their use in interpreting metabolic function from a pathway-oriented perspective. J. Theor. Biol. **248**, 203–229 (2000)
20. Wiback, J.S., Palsson, B.O.: Extreme pathway analysis of human red blood cell metabolism. Biophys. J . **83**, 808–818 (2003)
21. Barbehenn, M.: A note on the complexity of Dijkstra's algorithm for graphs with weighted vertices. IEEE Trans. Comput. **47**(2), 263 (1998)
22. Provost, A., Bastin, G.: Metabolic flux analysis: an approach for solving non-stationary undetermined systems. In: Troch, I., Breitenecker, F., Editors, 5th MATHMOD, pp. 5–10 (2006)

A Partially Decentralised Protocol for a Distributed Encrypted Storage System

Alessandro Languasco(✉)

Dipartimento di Matematica "Tullio Levi-Civita", Università di Padova, via Trieste 63, 35121 Padova, Italy
alessandro.languasco@unipd.it

Abstract. Decentralization is an important feature both for security and for reliability as it reduces the risks of collusion and impact of the single-point-of-failure vulnerability in a system. After having recalled the re-encryption procedure – mainly to establish notations – we will describe a modification of the distributed encrypted storage protocol by Ateniese-Fu-Green-Hohenberger. Such a protocol is a strongly centralized one since it assigns to the manager the possibility of storing all the symmetric keys involved in enciphering the resources required by the users of the system. Our goal here is to show how to reduce the role of the manager as being just a certifier of the access requests to the shared resources. We will achieve this by using standard public-key cryptography tools like the digital signature with message integrity.

Keyword: Re-encryption

1 Introduction: Some Remarks on the Re-Encryption Asymmetric Methods

We assume to work with an asymmetric cryptosystem and, for any user $\mathtt{X}$ of the system, we denote the enciphering function as $f_{\mathtt{X}}$, which depends on a public ket $p_{\mathtt{X}}$, and the corresponding deciphering functions as $f_{\mathtt{X}}^{-1}$, which depends on a private key $s_{\mathtt{X}}$.

We want to allow another user of the system $\mathtt{Y}$, $\mathtt{Y} \neq \mathtt{X}$, to access, after being authorized by $\mathtt{X}$, to a plaintext m originally encrypted for the used $\mathtt{X}$ only. In the protocol we will need a *trusted third party*, called *proxy* (PR), that stores the authorization $\mathtt{X}$ gave to PR to transfer to the user $\mathtt{Y}$ the messages originally sent to him.

So we need to find a way to let the user PR be able to perform a *re-encoding* procedure of an enciphered message $c_{\mathtt{X}} = f_{\mathtt{X}}(m)$, originally sent to $\mathtt{X}$, so that he will obtain $c_{\mathtt{Y}} = f_{\mathtt{Y}}(m)$, an enciphered version of m that can be deciphered only by the user $\mathtt{Y}$, different from the original receiver $\mathtt{X}$.

J. Bravo and G. Urzáiz (Eds.): UCAmI 2023, LNNS 841, pp. 195–204, 2023.
https://doi.org/10.1007/978-3-031-48590-9_19

Moreover, we need to ensure that the proxy PR will not be able to read the plaintext m during the re-encoding procedure.

We will show now the first solution proposed by Blaze-Bleumer-Strauss [2] in 1998 and then the one of Ateniese-Fu-Green-Hohenberger [1] which is the first one in which the re-encoding idea is applied with a sufficiently good security level. Both methods are similar to the classical ElGamal public key cryptosystem, see, e.g., Koblitz [6, Ch. IV-3].

In the following we will denote q as a prime number, G as a cyclic group of order q (we can also consider $G = (\mathbb{Z}/_{q\mathbb{Z}})^*$ with the modular product) and g as a generator of G, i.e., $\langle g \rangle = G$. We recall that this means that $G = \{g^x : x \in \{0, \dots, q-2\}\}$ and that the exponents of g have the additive group structure of $(\mathbb{Z}/_{(q-1)\mathbb{Z}}, +)$. As usual, we will denote the set of invertible elements of $(\mathbb{Z}/_{(q-1)\mathbb{Z}}, \cdot)$ as $(\mathbb{Z}/_{(q-1)\mathbb{Z}})^*$. All these parameters are public ones.

1.1 Blaze-Bleumer-Strauss: An Atomic Re-encryption System

The plaintexts space is $\mathfrak{M} = G = (\mathbb{Z}/_{q\mathbb{Z}})^*$, the ciphertexts one is $\mathfrak{C} = G \times G$, the secret keys space is $\mathfrak{K}_s = (\mathbb{Z}/_{(q-1)\mathbb{Z}})^*$ and the public keys space is $\mathfrak{K}_p = G$.

Let A, different from PR, be a generic user of the system and remark that $\mathfrak{M}_{\mathtt{A}} = \mathfrak{M}$, $\mathfrak{C}_{\mathtt{A}} = \mathfrak{C}$ for every A; the same happens for $\mathfrak{K}_s$ and $\mathfrak{K}_p$ too. Such a user A has to

1) randomly generate a private key $s_{\mathtt{A}} \in \mathfrak{K}_s$; remark that $s_{\mathtt{A}}$ is an invertible element of $\mathbb{Z}/_{(q-1)\mathbb{Z}}$, i.e., $(s_{\mathtt{A}}, q-1) = 1$;
2) compute his public key $p_{\mathtt{A}} \in \mathfrak{K}_p$, $p_{\mathtt{A}} \equiv g^{s_{\mathtt{A}}} \pmod{q}$ (or, in other words, $p_{\mathtt{A}} = g^{s_{\mathtt{A}}}$ in G).

The *enciphering function* of a generic user A, $f_{\mathtt{A}} : \mathfrak{M}_{\mathtt{A}} \longrightarrow \mathfrak{C}_{\mathtt{A}}$, $\mathfrak{M}_{\mathtt{A}} = G$, $\mathfrak{C}_{\mathtt{A}} = G \times G$, works as follows. A user C that wants to send a plaintext $m \in G$ to A has to:

1) randomly generate $r \in \mathbb{Z}/_{(q-1)\mathbb{Z}}$;
2) compute the following three quantities: $g^r \pmod{q}$, $g^r m \pmod{q}$ e $p_{\mathtt{A}}^r \pmod{q}$;
3) send to A the ciphertext $c_{\mathtt{A}} := (g^r m, p_{\mathtt{A}}^r)$[1].

The *deciphering function* of A, $f_{\mathtt{A}}^{-1} : \mathfrak{C}_{\mathtt{A}} \longrightarrow \mathfrak{M}_{\mathtt{A}}$, is defined as follows. After having received from C the ciphertext $c_{\mathtt{A}} \in G \times G$, a user A has to:

1) use his private key $s_{\mathtt{A}} \in \mathfrak{K}_s$ on the second part of $c_{\mathtt{A}}$; hence A must
   a) compute $s_{\mathtt{A}}^{-1} \pmod{q-1}$;
   b) compute $(p_{\mathtt{A}}^r)^{s_{\mathtt{A}}^{-1}} \pmod{q}$; in this way A obtains $p_{\mathtt{A}}^{r s_{\mathtt{A}}^{-1}} \equiv g^{r s_{\mathtt{A}} s_{\mathtt{A}}^{-1}} \equiv g^r \pmod{q}$;
   c) compute the inverse of $g^r \pmod{q}$, i.e., $(g^r)^{-1} \equiv g^{-r} \pmod{q}$.

[1] Remark that in the ElGamal cryptosystem the plaintext m would be enciphered as $(p_{\mathtt{A}}^r m, g^r)$ thus letting the blinding factor depend from A's public key; in the Blaze-Bleumer-Strauss system the first part of $c_{\mathtt{A}}$ does not depend on A.

2) perform the product in G between the first part of $c_{\mathtt{A}}$ and the quantity just obtained; he gets $(g^r m)g^{-r} \equiv g^r g^{-r} m \equiv g^r (g^r)^{-1} m \equiv m \pmod{q}$, thus completely deciphering $c_{\mathtt{A}}$.

The special user PR, the proxy of the system, can use a *re-encryption* function $f_{\mathtt{A}\to\mathtt{B}}\colon \mathfrak{C}_{\mathtt{A}} \to \mathfrak{C}_{\mathtt{B}}$, $\mathfrak{C}_{\mathtt{A}} = \mathfrak{C}_{\mathtt{B}} = G \times G$, between the ciphertexts spaces of A and B. Such a function makes use of a *re-encryption key* $\mathtt{rk}_{\mathtt{A}\to\mathtt{B}} := s_{\mathtt{A}}^{-1} s_{\mathtt{B}} \in \mathfrak{K}_s$, i.e., $\mathtt{rk}_{\mathtt{A}\to\mathtt{B}} \equiv s_{\mathtt{A}}^{-1} s_{\mathtt{B}} \pmod{q-1}$. The re-encryption function is defined as follows. Assume that A received from C the ciphertext $c_{\mathtt{A}} = (g^r m, p_{\mathtt{A}}^r) \in G \times G$; to re-encipher $c_{\mathtt{A}}$ in $c_{\mathtt{B}}$, PR has to use the re-encryption key $\mathtt{rk}_{\mathtt{A}\to\mathtt{B}}$ on the second part of $c_{\mathtt{A}}$. Hence the special user PR performs the following operations:

1) computes $(p_{\mathtt{A}}^r)^{\mathtt{rk}_{\mathtt{A}\to\mathtt{B}}} \pmod{q}$; thus getting $p_{\mathtt{A}}^{r s_{\mathtt{A}}^{-1} s_{\mathtt{B}}} \equiv g^{r s_{\mathtt{A}} s_{\mathtt{A}}^{-1} s_{\mathtt{B}}} \equiv g^{r s_{\mathtt{B}}} \pmod{q} \equiv p_{\mathtt{B}}^r \pmod{q}$;
2) combines the first part of $c_{\mathtt{A}}$ with the quantity just computed thus obtaining $c_{\mathtt{B}} = (g^r m, p_{\mathtt{B}}^r) \in G \times G$;
3) sends $c_{\mathtt{B}}$ to B.

The user B, after having received $c_{\mathtt{B}}$, can decipher and read the plaintext m originally sent to A from C.

The Blaze-Bleumer-Strauss system has several problems. First, it is important to remark that $\mathtt{rk}_{\mathtt{A}\to\mathtt{B}}$ is computed by using both the secret keys of A and B; hence such users must collaborate between them (and with PR) to build such a re-encryption key *before* being able to start to communicate using this system: so they have to overcome a *pre-sharing* problem. Moreover, if A and B, with the goal of generate $\mathtt{rk}_{\mathtt{A}\to\mathtt{B}}$, both send to PR their secret keys as plaintexts, it is clear that PR will know both of them! And this violates one of the basic requirement of the public key cryptography. One can try to solve this problem by using some *secret splitting* or *secret sharing* protocols to let PR obtain $\mathtt{rk}_{\mathtt{A}\to\mathtt{B}}$ without knowing its two factors. In any case this does not solve the pre-sharing problem between A, B and PR.

We further remark that $(\mathtt{rk}_{\mathtt{A}\to\mathtt{B}})^{-1} = (s_{\mathtt{A}}^{-1} s_{\mathtt{B}})^{-1} = (s_{\mathtt{A}}^{-1})^{-1} s_{\mathtt{B}}^{-1} = s_{\mathtt{B}}^{-1} s_{\mathtt{A}} = \mathtt{rk}_{\mathtt{B}\to\mathtt{A}}$ in $(\mathbb{Z}/_{(q-1)\mathbb{Z}})^*$. Hence, if A allows B to read his messages, as an immediate consequence we get that any ciperthext originally sent to B can be automatically re-encoded for A! And B has no way to prevent this from happening. This means that the Blaze-Bleumer-Strauss re-encryption system is a fully *bidirectional* one.

Finally, if PR and A are dishonest users, they can collaborate to gain the knowledge of B's secret key $s_{\mathtt{B}}$ (also in the case B used a secret sharing protocol to communicate it to PR). In fact, in this case PR can compute the modular product $s_{\mathtt{A}} \mathtt{rk}_{\mathtt{A}\to\mathtt{B}} \equiv s_{\mathtt{A}} s_{\mathtt{A}}^{-1} s_{\mathtt{B}} \equiv s_{\mathtt{B}} \pmod{q-1}$ thus obtaining $s_{\mathtt{B}}$!

1.2 The Unidirectional Method by Ateniese-Fu-Green-Hohenberger

In 2006 Ateniese-Fu-Green-Hohenberger [1] proposed the following *unidirectional* re-encryption procedure. The main idea is to link the original discrete-log problem to another one which depends on the first (that, we recall, is very similar to the classical ElGamal method) but works on a different cyclic group. So we need a way to perform this translation from the first problem to the second and this, from a mathematical point of view, can be achieved by using a *bilinear* map.

Definition 1. *Let q be a prime number and let G_1, G_2 be two cyclic groups both having the same order q. We will say that the function $\mathbb{e}\colon G_1 \times G_1 \to G_2$ is a* bilinear map *if and only if for every $x_1, x_2 \in G_1$ and for every $a, b \in \mathbb{Z}/_{(q-1)\mathbb{Z}}$ we have that $\mathbb{e}(x_1^a, x_2^b) = \mathbb{e}(x_1, x_2)^{ab}$. Moreover, if g is a generator of G_1, we will say the $\mathbb{e}$ is* non-degenerate *if and only if $z := \mathbb{e}(g, g)$ is a generator of G_2.*

In this paper we will not consider the problem of proving that such bilinear maps exist; one can find several examples of bilinear maps that can be used in these applications in [4].

In the following, we assume that $G_1 = \langle g \rangle$ and $G_2 = \langle \mathbb{e}(g, g) \rangle$ (in other words, that $\mathbb{e}$ is bilinear and non-degenerate) are both cyclic groups of order q, q being a prime number. Moreover, we assume that q, g and $\mathbb{e}$ are of public knowledge and that $\mathbb{e}$ can be efficiently computed by every user of the cryptosystem. Clearly, it immediately follow that $\mathbb{e}(g, g)$ is public knowledge too.

We now describe the key-generation step in the Ateniese-Fu-Green-Hohenberger protocol.

Let $\mathtt{X}$ be a user of the cryptosystem. He randomly generates a secret key $s_{\mathtt{X}} \in (\mathbb{Z}/_{(q-1)\mathbb{Z}})^*$ and compute his public key as $p_{\mathtt{X}} := g^{s_{\mathtt{X}}}$ in G_1, or, in other words, $p_{\mathtt{X}} \equiv g^{s_{\mathtt{X}}} \pmod q$. Moreover, if $\mathtt{X}$ and $\mathtt{Y}$ are two users, $\mathtt{X} \neq \mathtt{Y}$, the re-encryption key $\mathtt{rk}_{\mathtt{X}\to\mathtt{Y}}$ from $\mathtt{X}$ to $\mathtt{Y}$ is defined as $\mathtt{rk}_{\mathtt{X}\to\mathtt{Y}} := g^{s_{\mathtt{Y}} s_{\mathtt{X}}^{-1}} = p_{\mathtt{Y}}^{s_{\mathtt{X}}^{-1}} \in G_1$. We immediately remark that $\mathtt{X}$ can compute $\mathtt{rk}_{\mathtt{X}\to\mathtt{Y}}$ because $p_{\mathtt{Y}}$ is public knowledge and, using $s_{\mathtt{X}}$, he can use the Extended Euclidean algorithm to efficently obtain $s_{\mathtt{X}}^{-1}$.

An important feature of this protocol is that we can distinguish if a ciphertext comes from a re-encryption procedure or not: a *first level encoding* is the procedure in which a ciphertext to be sent to the user $\mathtt{X}$ is obtained by using the re-encryption key; hence the final result is $c'_{\mathtt{X}} \in G_2 \times G_2$. A *second level encoding* is the usual one (the re-encryption key is not used at all); hence it produces $c''_{\mathtt{X}} \in G_2 \times G_1$.

We now assume that $\mathtt{A}, \mathtt{B}, \mathtt{C}$ are three different users of the cryptosystem.

The *second level enciphering function* for the user $\mathtt{A}$, $f''_{\mathtt{A}}\colon \mathfrak{M}_{\mathtt{A}} \to \mathfrak{C}''_{\mathtt{A}}$, has $\mathfrak{M}_{\mathtt{A}} := G_2$, $\mathfrak{C}''_{\mathtt{A}} = G_2 \times G_1$, $\mathfrak{K}_{p,\mathtt{A}} := G_1$, $\mathfrak{K}_{s,\mathtt{A}} := (\mathbb{Z}/_{(q-1)\mathbb{Z}})^*$. It is a public function, in the sense that it can be computed by any user of the cryptosystem that wants to send a message to $\mathtt{A}$.

Assume that the user $\mathtt{C}$ wishes to send the plaintext $m \in G_2$ to $\mathtt{A}$. The user $\mathtt{C}$ has to:

1) randomly generate a parameter $r \in \mathbb{Z}/_{(q-1)\mathbb{Z}}$;
2) compute z^r in G_2, $z^r m$ in G_2 and $p_{\mathtt{A}}^r \in G_1$;
3) compute $c''_{\mathtt{A}} := (z^r m, p_{\mathtt{A}}^r)$ and sent it to $\mathtt{A}$.[2]

[2] As in the Blaze-Bleumer-Strauss algorithm, the blinding factor that is used to obscure m works only on the first part of $c''_{\mathtt{A}}$ and does not depend on the user $\mathtt{A}$.

The *second level deciphering function* of the receiver A, $(f''_{\mathtt{A}})^{-1}\colon \mathfrak{C}''_{\mathtt{A}} \to \mathfrak{M}_{\mathtt{A}}$, works as follows. Assume that A received the cipertext $c''_{\mathtt{A}} := (z^r m, p^r_{\mathtt{A}})$. To decipher it, the user A has to

1) compute, starting from the knowledge of his secret key $s_{\mathtt{A}} \in (\mathbb{Z}/_{(q-1)\mathbb{Z}})^*$, the quantity $s_{\mathtt{A}}^{-1}$;
2) compute $g^{s_{\mathtt{A}}^{-1}} \bmod q$;
3) use the second part of $c''_{\mathtt{A}}$ to compute $\mathbb{e}(p^r_{\mathtt{A}}, g^{s_{\mathtt{A}}^{-1}}) = \mathbb{e}(g^{s_{\mathtt{A}} r}, g^{s_{\mathtt{A}}^{-1}}) = \mathbb{e}(g,g)^{s_{\mathtt{A}} r s_{\mathtt{A}}^{-1}} = z^r \in G_2$;
4) compute the inverse of z^r in G_2; thus obtaining $(z^r)^{-1} = z^{-r}$ in G_2;
5) use the first part of $c''_{\mathtt{A}}$ to compute $(z^r m)z^{-r} = m$ in G_2.

The special user PR of the system can also use the *re-encryption function* which is defined in the following way: $f'_{\mathtt{A}\to\mathtt{B}}\colon \mathfrak{C}''_{\mathtt{A}} \to \mathfrak{C}'_{\mathtt{B}}$ has $\mathfrak{C}''_{\mathtt{A}} = G_2 \times G_1$, $\mathfrak{C}'_{\mathtt{B}} = G_2 \times G_2$ and $\mathfrak{K}_{\mathtt{rk}_{\mathtt{A}\to\mathtt{B}}} = G_1$. It is a function that *only* PR can compute.

Assume that the user C sent to A the second level ciphertext $c''_{\mathtt{A}} = (z^r m, p^r_{\mathtt{A}})$ and that A previously allowed the proxy user PR to re-encode his second level ciphertexts to the user B by using the re-encryption key $\mathtt{rk}_{\mathtt{A}\to\mathtt{B}} = p_{\mathtt{B}}^{s_{\mathtt{A}}^{-1}} = g^{s_{\mathtt{B}} s_{\mathtt{A}}^{-1}} \in G_1$. Hence PR computes the re-encryption function $f'_{\mathtt{A}\to\mathtt{B}}$ by

1) using the second part of $c''_{\mathtt{A}}$ thus obtaining that $\mathbb{e}(\mathtt{rk}_{\mathtt{A}\to\mathtt{B}}, p^r_{\mathtt{A}}) = \mathbb{e}(g^{s_{\mathtt{B}} s_{\mathtt{A}}^{-1}}, g^{s_{\mathtt{A}} r}) = \mathbb{e}(g,g)^{s_{\mathtt{B}} s_{\mathtt{A}}^{-1} s_{\mathtt{A}} r} = \mathbb{e}(g,g)^{s_{\mathtt{B}} r} = z^{s_{\mathtt{B}} r}$ in G_2;
2) pairing the first part of $c''_{\mathtt{A}}$ and the quantity just obtained thus getting the first level ciphertext $c'_{\mathtt{B}} = (z^r m, z^{s_{\mathtt{B}} r}) \in G_2 \times G_2$;
3) sending to B the quantity $c'_{\mathtt{B}}$.[3]

Now B receives from PR the quantity $c'_{\mathtt{B}} \in G_2 \times G_2$ and he can recognise that it is a first level encoded message. Hence, to decipher, B has to use the function $(f'_{\mathtt{B}})^{-1} : \mathfrak{C}'_{\mathtt{B}} \to \mathfrak{M}_{\mathtt{B}}$, where $\mathfrak{M}_{\mathtt{B}} = G_2$ e $\mathfrak{C}'_{\mathtt{B}} = G_2 \times G_2$. Such a function works as follows:

1) B computes in $(\mathbb{Z}/_{(q-1)\mathbb{Z}})^*$ the inverse of his own private key $s_{\mathtt{B}}$ thus getting $s_{\mathtt{B}}^{-1}$;
2) B uses the second part of $c'_{\mathtt{B}}$ to obtain $(z^{s_{\mathtt{B}} r})^{s_{\mathtt{B}}^{-1}} = z^{s_{\mathtt{B}} r s_{\mathtt{B}}^{-1}} = z^r$ in G_2;
3) B computes the inverse of z^r in G_2;
4) B uses the first part of $c'_{\mathtt{B}}$ to obtain $z^r m z^{-r} = m$ in G_2.

In this way B is able to read the plaintext m that was sent from C to A first, and re-encoded by PR.

It is clear that the security of the algorithm just described depends on the computational hardness of solving the discrete logarithm problem both in G_1 and in G_2. And that the unidirectionality of the procedure depends on the hardness of solving the discrete logarithm problem in G_1: in fact PR, without solving

[3] Remark that there is no way to detect that $c'_{\mathtt{B}}$ was obtained starting from $c''_{\mathtt{A}}$; in other words, an intruder, or the user B, is not able to understand to whom the initial part of $c'_{\mathtt{B}}$ was first sent.

a suitable instance of this problem, cannot obtain any information on $\mathtt{rk}_{\mathtt{B}\to\mathtt{A}} = p_{\mathtt{A}}^{s_{\mathtt{B}}^{-1}} = g^{s_{\mathtt{A}}s_{\mathtt{B}}^{-1}} \in G_1$ from the knowledge of $\mathtt{rk}_{\mathtt{A}\to\mathtt{B}} = p_{\mathtt{B}}^{s_{\mathtt{A}}^{-1}} = g^{s_{\mathtt{B}}s_{\mathtt{A}}^{-1}} \in G_1$. On the other hand, it is clear that every intruder that is able to solve the discrete logarithm problem both in G_1 and G_2 can perform both the first and the second level decryption procedures and he is also able to re-encode messages (thus illegally acting as a PR).

We remark that no pre-sharing procedure between A and B is needed to compute $\mathtt{rk}_{\mathtt{A}\to\mathtt{B}}$ and $\mathtt{rk}_{\mathtt{B}\to\mathtt{A}}$; moreover, no secret sharing or secret splitting procedure is needed too.

We also remark that a first level encoding $c'_{\mathtt{B}}$ (hence the output of a first level enciphering function) cannot be further re-encoded: this depends on the fact that $c'_{\mathtt{B}} \in G_2 \times G_2$ while, to be further re-encoded, it should be an element of $G_2 \times G_1$.

It is important to remark that a dishonest proxy PR could agree with B (or in the case in which B illegally impersonates PR), and further re-encode to another user D a message m sent from C to A and re-encoded by PR to B. If $c'_{\mathtt{B}} = (z^r m, z^{s_{\mathtt{B}}r})$ is the first level ciphertext sent to B by PR, and if B shares the knowledge of $s_{\mathtt{B}}^{-1}$ with PR, then PR can compute the plaintext m. Now PR can randomly generate $r_1 \in \mathbb{Z}/_{(q-1)\mathbb{Z}}$ and obtain $z^{r_1}m \in G_2$. And, exploiting the fact that both $p_{\mathtt{B}}$ and $p_{\mathtt{D}}$ are public knowledge, PR can compute (recall that PR knows what $s_{\mathtt{B}}^{-1}$ is):

$$\mathbb{e}(p_{\mathtt{B}}^{r_1}, p_{\mathtt{D}}^{s_{\mathtt{B}}^{-1}}) = \mathbb{e}(g^{s_{\mathtt{B}}r_1}, g^{s_{\mathtt{D}}s_{\mathtt{B}}^{-1}}) = \mathbb{e}(g,g)^{s_{\mathtt{B}}r_1 s_{\mathtt{D}}s_{\mathtt{B}}^{-1}} = \mathbb{e}(g,g)^{r_1 s_{\mathtt{D}}} = z^{r_1 s_{\mathtt{D}}} \in G_2.$$

Hence PR, by combining $z^{r_1}m$ with the quantity just computed, can form the pair $(z^{r_1}m, z^{r_1 s_{\mathtt{D}}})$ which is a correct first level encryption that can be send from PR to D. In conclusion, if PR illegally agrees with B, then PR can read the plaintext m and he can also send $c'_{\mathtt{D}} = (z^{r_1}m, z^{r_1 s_{\mathtt{D}}})$ to the user D thus passing the plaintext m, originally sent from C to A, to D *without having any authorisation from the users A and C*. Moreover, D has no clue to understand who was the original sender of m.

We finally recall that Ateniese-Fu-Green-Hohenberger in [1] proposed a variant of their method in which every user has to generate a pair of secret keys (one secret exponent for the generator g of G_1 and another one for the generator z of G_2) so that any user can express his consensus to any re-encoding procedure of the received messages. In this case too the blinding factor used to obscure the plaintext m in the first level encoding procedure does not depend from any user-related object; hence our previous remark about the effect that an illegal connection between PR and B can have on the global correctness of the re-encoding procedure still stands.

2 Distributed Encrypted Storage

In this section we first recall the Ateniese-Fu-Green-Hohenberger (AFGH) Distributed Encrypted Storage protocol based on the re-encryption idea and then we will present an original one that mitigates the main problems present in AFGH.

2.1 The Ateniese-Fu-Green-Hohenberger Protocol

We now describe the Ateniese-Fu-Green-Hohenberger protocol, see [1], to operate a Distributed Encrypted Storage system.

It is a *centralised* protocol in which there exists a special user MG, the *manager* of the system. We will use two different cryptosystems. The first is a symmetric one having a sufficiently large number of keys k (AES, for instance, see [3]), while the second is an asymmetric one in which a suitable re-encryption procedure is also available. We further assume that a *key server* K is available to store the keys the manager MG will send to the users of the system.

For every user X of the system we will denote the enciphering function of the public key cryptosystem as $f_{\mathtt{X}}$ (it is public in the sense that it depends from a public key $p_{\mathtt{X}}$) and with $f_{\mathtt{X}}^{-1}$ we will denote the corresponding deciphering function (which is private, in the sense that it depends on a secret key $s_{\mathtt{X}}$). Moreover, we will denote as $\mathtt{rk}_{\mathtt{X}\to\mathtt{Y}}$ the *re-encryption* key the manager MG, that is playing the role of PR[4], will use to transform $f_{\mathtt{X}}(m)$ into $f_{\mathtt{Y}}(m)$, where m is a plaintext and $\mathtt{X}, \mathtt{Y}, \mathtt{X} \neq \mathtt{Y}$ are two users.

Finally, denote as $\mathscr{D}$ the data storing device we are using and as F a file stored on $\mathscr{D}$.

The Ateniese-Fu-Green-Hohenberger Distributed Encrypted Storage protocol works as follows.

1) MG enciphers every file F stored onto $\mathscr{D}$ using a symmetric key $k_{\mathtt{F}}$ (for example, $k_{\mathtt{F}}$ is an AES key); it is important that $k_{\mathtt{F}_1} \neq k_{\mathtt{F}_2}$ if $\mathtt{F}_1 \neq \mathtt{F}_2$.
2) MG enciphers using his public enciphering function $f_{\mathtt{MG}}$ the symmetric keys previously used; he obtains the set $\{f_{\mathtt{MG}}(k_{\mathtt{F}_j})\colon \text{for every } \mathtt{F}_j \in \mathscr{D}\}$. MG sends such a set to the key server K.[5]
3) A user B, different from MG, asks to MG to access the file $\mathtt{F} \in \mathscr{D}$. In this case MG generates the re-encryption key[6] $\mathtt{rk}_{\mathtt{MG}\to\mathtt{B}}$ and he sends it to K. K stores $\mathtt{rk}_{\mathtt{MG}\to\mathtt{B}}$ so that it can be re-used by B if he will later ask to access to another file stored on $\mathscr{D}$.
4) The key server K uses $\mathtt{rk}_{\mathtt{MG}\to\mathtt{B}}$ onto $f_{\mathtt{MG}}(k_{\mathtt{F}})$ thus obtaining $f_{\mathtt{B}}(k_{\mathtt{F}})$. Then K sends $f_{\mathtt{B}}(k_{\mathtt{F}})$ to B[7].
5) B can now decipher the last message using $f_{\mathtt{B}}^{-1}$ thus getting $f_{\mathtt{B}}^{-1}(f_{\mathtt{B}}(k_{\mathtt{F}})) = k_{\mathtt{F}}$; he now knows the symmetric key $k_{\mathtt{F}}$ that MG used to encipher the file F. Using $k_{\mathtt{F}}$, B can now access to F.

[4] In fact, starting from the second request, it will be the key server K to act as the PR user of the re-encryption protocol.

[5] Remark that MG can store a copy of the simmetric keys $k_{\mathtt{F}_j}$.

[6] Since MG knows his own private key $s_{\mathtt{MG}}$, we can get rid of the re-encryption system: MG can compute directly $p_{\mathtt{B}} s_{\mathtt{MG}}$ to obtain $\mathtt{rk}_{\mathtt{MG}\to\mathtt{B}}$. In fact, it is only from B's second request that the key server K acts as PR, since we do not need to compute again $\mathtt{rk}_{\mathtt{MG}\to\mathtt{B}}$.

[7] From his second request on, B can directly ask to the key server K to access to a file F stored on $\mathscr{D}$.

It is clear that MG, having the possibility of storing all the symmetric keys $k_{\mathtt{F}}$, has a strongly privileged role that centralises the system architecture. This also introduces a single-point-of-failure factor of vulnerability in this system. Moreover, it is also clear that an identification protocol should be used to certify the requests that any user X can send to MG or to K.

2.2 A Partially Decentralised Protocol with User Identification and Data Integrity Verification

We will continue to use the notations and definitions of the previous subsection. We will also denote a public *hash* function as h, a *nickname* of the user X as $\mathrm{id}_{\mathtt{X}}$ (stored together with X's public key) and with $d_{\mathtt{F}}$ a plaintext description of the file $\mathtt{F} \in \mathscr{D}$. The classical concepts of *digital signature* and *hash* function can be found, e.g., in [5, ch. 7 and 11]. Our protocol runs as follows.

To commit his file $\mathtt{F}_{\mathtt{X}}$ to the storing device $\mathscr{D}$, the user X:

X-1) enciphers $\mathtt{F}_{\mathtt{X}}$ using a key $k_{\mathtt{F}_{\mathtt{X}}}$ of a symmetric cryptosystem (for instance, assume that $k_{\mathtt{F}_{\mathtt{X}}}$ is an AES-key); it is important that $k_{\mathtt{F}_{1,\mathtt{X}}} \neq k_{\mathtt{F}_{2,\mathtt{X}}}$ if $\mathtt{F}_{1,\mathtt{X}} \neq \mathtt{F}_{2,\mathtt{X}}$;

X-2) enciphers such a symmetric key with his (public) enciphering function thus obtaining $f_{\mathtt{X}}(k_{\mathtt{F}_{\mathtt{X}}})$;

X-3) computes $h(k_{\mathtt{F}_{\mathtt{X}}})$ (the hash image of $k_{\mathtt{F}_{\mathtt{X}}}$), $h(\mathrm{id}_{\mathtt{X}})$ (the hash image of $\mathrm{id}_{\mathtt{X}}$) and $h(\mathtt{F}_{\mathtt{X}})$ (the hash image of $\mathtt{F}_{\mathtt{X}}$); he then concatenates them into $[h(k_{\mathtt{F}_{\mathtt{X}}}) \mid h(\mathrm{id}_{\mathtt{X}}) \mid h(\mathtt{F}_{\mathtt{X}})]$;

X-4) signs such a quantity by using his (secret) deciphering function. He obtains the quantiity $f_{\mathtt{X}}^{-1}([h(k_{\mathtt{F}_{\mathtt{X}}}) \mid h(\mathrm{id}_{\mathtt{X}}) \mid h(\mathtt{F}_{\mathtt{X}})])$;

X-5) concatenates the previous result with $f_{\mathtt{X}}(k_{\mathtt{F}_{\mathtt{X}}})$; the result is

$$f_{\mathtt{X}}^{-1}([h(k_{\mathtt{F}_{\mathtt{X}}}) \mid h(\mathrm{id}_{\mathtt{X}}) \mid h(\mathtt{F}_{\mathtt{X}})]) \Big| f_{\mathtt{X}}(k_{\mathtt{F}_{\mathtt{X}}}).$$

X-6) He then writes a short plaintext $d_{\mathtt{F}_{\mathtt{X}}}$ to describe $\mathtt{F}_{\mathtt{X}}$.

X-7) X sends then to MG, using the (public) enciphering function of MG, the latest quantity previously computed; in other words X sents to MG the quantity:

$$f_{\mathtt{MG}}\Big(f_{\mathtt{X}}^{-1}([h(k_{\mathtt{F}_{\mathtt{X}}}) \mid h(\mathrm{id}_{\mathtt{X}}) \mid h(\mathtt{F}_{\mathtt{X}})]) \Big| f_{\mathtt{X}}(k_{\mathtt{F}_{\mathtt{X}}}) \Big| d_{\mathtt{F}_{\mathtt{X}}}\Big).$$

This ends the committing part of the procedure X has to perform. It is now MG that has to complete his part of the protocol:

MG-1) MG verifies X's identity using $h(\mathrm{id}_{\mathtt{X}})$ contained in the first signed part of the quantity sent in point X-7 before;

MG-1) MG publishes on his server that X notified him the fact that he owns a file described by $d_{\mathtt{F}_{\mathtt{X}}}$. To do so, MG writes on his server a signed version of this quantity, i.e., $f_{\mathtt{MG}}^{-1}(\mathrm{id}_{\mathtt{X}} \mid d_{\mathtt{F}_{\mathtt{X}}})$.[8]

[8] Remark that MG can access $d_{\mathtt{F}_{\mathtt{X}}}$ from the quantity X sent to him. Moreover, he can also verify X's identity by comparing the signed $h(\mathrm{id}_{\mathtt{X}})$ quantity X sent before with the analogous one he can independently compute.

Now we see how a user Y, $\mathtt{Y} \neq \mathtt{X}$, can access to $\mathtt{F_X}$. The user Y reads the $\mathtt{F_X}$-description stored by MG and asks him to access to $\mathtt{F_X}$. To do so

Y-1) Y sends to MG the quantity $f_{\mathtt{MG}}(f_{\mathtt{Y}}^{-1}(\mathrm{id}_{\mathtt{Y}} \,|\, \mathrm{id}_{\mathtt{X}} \,| d_{\mathtt{F_X}}))$.
Y-2) MG, after having verified Y's identity, generates the re-encryption key[9] $\mathtt{rk}_{\mathtt{X}\to\mathtt{Y}}$ and he sends it to the key server K.
Y-3) The key server K stores $\mathtt{rk}_{\mathtt{X}\to\mathtt{Y}}$ (so that it can be re-used if in the future Y will ask for another file owned by X and stored onto $\mathscr{D}$).
Y-4) MG deciphers X's message (sent in point X-7 before) and, using $\mathtt{rk}_{\mathtt{X}\to\mathtt{Y}}$ on the second part, obtains $f_{\mathtt{Y}}(k_{\mathtt{F_X}}) = \mathtt{rk}_{\mathtt{X}\to\mathtt{Y}}(f_{\mathtt{X}}(k_{\mathtt{F_X}}))$; the first part of X's message (sent in point X-7 before) is left untouched.
Y-5) MG signs the first part of X's message (sent in point X-7 before) and the quantity $f_{\mathtt{Y}}(k_{\mathtt{F_X}})$ just obtained using the re-encryption key. He then concatenates and signs the nickname of X. The final result is

$$f_{\mathtt{MG}}^{-1}\left(f_{\mathtt{X}}^{-1}([h(k_{\mathtt{F_X}}) \,|\, h(\mathrm{id}_{\mathtt{X}}) \,|\, h(\mathtt{F_X})])\right) \Big| f_{\mathtt{MG}}^{-1}(\mathrm{id}_{\mathtt{X}}) \Big| f_{\mathtt{MG}}^{-1}(f_{\mathtt{Y}}(k_{\mathtt{F_X}}))$$

that he sends to Y after having further applied $f_{\mathtt{Y}}$.
Y-6) The user Y can now decipher the last message using $f_{\mathtt{Y}}^{-1}$; he obtains the following quantities (all signed by MG):

a) $f_{\mathtt{X}}^{-1}([h(k_{\mathtt{F_X}}) \,|\, h(\mathrm{id}_{\mathtt{X}}) \,|\, h(\mathtt{F_X})])$, i.e., $h(k_{\mathtt{F_X}}) \,|\, h(\mathrm{id}_{\mathtt{X}}) \,|\, h(\mathtt{F_X})$ all signed by X;
b) $f_{\mathtt{Y}}(k_{\mathtt{F_X}})$, i.e., further using $f_{\mathtt{Y}}^{-1}$, the symmetric key $k_{\mathtt{F_X}}$ needed to access the file $\mathtt{F_X}$ committed by X;
c) an identification of X signed by MG.

Hence, at the end of all the protocol, the user Y can:

i) verify $\mathrm{id}_{\mathtt{X}}$ signed by MG, by computing its hash-image and comparing it with the quantity $h(\mathrm{id}_{\mathtt{X}})$, computed and signed by X and further signed by MG;
ii) verify the correctness of $k_{\mathtt{F_X}}$ by computing its hash-image and comparing it with $h(k_{\mathtt{F_X}})$, computed and signed by X and further signed by MG;
iii) access to $\mathscr{D}$ and use $k_{\mathtt{F_X}}$ to decipher and obtain the file $\mathtt{F_X}$ committed by X;
iv) verify that the whole protocol is correct by computing the hash-image of $\mathtt{F_X}$ and comparing it with $h(\mathtt{F_X})$, computed and signed by X and further signed by MG.

We still have that MG and the key server K play an important role but in this case they are just certifying the committed files and the access requests to the shared resources. MG can just store enciphered or hashed quantities and the key server K is used to just store the re-encryption keys.

We finally remark that in every step of our protocol we used asymmetric encryption procedures and that every commit operation or access request must be correctly signed by the involved users X, Y, MG. Moreover, a data integrity verification is inserted too by using suitable hash images of the involved quantities. Clearly, in case at least one of such identity or data integrity verification will fail, the final user Y, or, in same cases, the manager MG, can reject the received data thus dropping the whole protocol execution.

[9] It is in this point that MG plays the role of the PR user of the cryptosystem we are using to re-encode messages.

3 Conclusion

We analysed some of the re-encryption protocols available and, to move toward a decentralized version of the Ateniese-Fu-Green-Hohenberger protocol for a distributed encrypted storage system, we proposed an enhanced version that leverages digital signature and hash-image primitives. Our version is less centralized than the original one since the manager MG of the system does not have anymore the possibility of storing all the used keys, but it only has the role of certifying the access requests to the shared resources. Besides, our version also allows for a data integrity verification procedure. On the other hand, our protocol has an increased complexity and the extensive use of the digital signature and hash images primitives increases its computational cost. In future work we will study how to reduce such a computational cost.

References

1. Ateniese, G., Fu, K., Green, M., Hohenberger, S.: Improved proxy reencryption schemes with applications to secure distributed storage. ACM Trans. Inf. Syst. Secur. **9**(1), 1–30 (2006)
2. Blaze, M., Bleumer, G., Strauss, M.: Divertible protocols and atomic proxy cryptography. In: Nyberg, K. (ed.) EUROCRYPT 1998. LNCS, vol. 1403, pp. 127–144. Springer, Heidelberg (1998). https://doi.org/10.1007/BFb0054122
3. Daemen, J., Rijmen, V.: The design of Rijndael, Springer, Berlin, Heidelberg (2002). https://doi.org/10.1007/978-3-662-04722-4
4. El Mrabet, N., Joye, M.: Guide to Pairing-Based Cryptography, first ed., Chapman and Hall/CRC (2017)
5. Knopse, H.: A course in cryptography. Am. Math. Soc. (2019)
6. Koblitz, N.: A course in number theory and cryptography, second ed., Springer-Verlag, (1994). https://doi.org/10.1007/978-1-4419-8592-7

Cloud-Native Application Security Training and Testing with Cyber Ranges

Enrico Russo[1], Giacomo Longo[1], Meriem Guerar[1], and Alessio Merlo[2](✉)

[1] DIBRIS - University of Genoa, Genova, Italy
{Enrico.Russo,Giacomo.Longo,Meriem.Guerar}@dibris.unige.it
[2] CASD - School of Advanced Defense Studies, Rome, Italy
alessio.merlo@ieee.org

Abstract. As cloud technology has become increasingly predominant in the last decade, more and more companies have been choosing to migrate to the cloud to leverage its cost-efficient services. Due to the hectic market pace, cloud security is often overlooked, thus leading to critical cyber attacks that can result in severe impacts, e.g., massive data leaks. Therefore, training appropriate personnel to secure cloud-native applications against these newly emerging threats is necessary. Currently, among the different cloud security training projects available, no environment is completely safe and gives full legal freedom since public providers host them, incurring their limitations. The proposed work aims to fill such a gap, discussing the implementation of a toolkit that can be used to implement a local cyber range safe and legally free from cloud providers' constraints that can host vulnerable cloud-native applications to create training scenarios. The said toolkit was used to host our vulnerable-by-design cloud-native application. It was successively administered to a class of students through a CTF competition to assess its educative potential.

Keywords: Cyber Ranges · Cloud-Native · Cloud Security · Security Training and Testing

1 Introduction

Cloud Native Applications (CNAs) is a well-established standard [15] adopted by most industry giants, such as Netflix [20]. CNAs can grant high resilience and adaptability to varying loads and faster delivery due to Continuous Integration/Continuous Delivery (CI/CD) pipelines. The positive cloud trend also involved governments migrating their public systems to the cloud to increase the efficiency and availability of their services [18].

Since CNAs are thriving and more sensitive data is being migrated to the cloud, security is fundamental, given that traditional security measures may be insufficient. CNAs combine classical vulnerabilities from the monolithic approach with new, cloud-specific ones [25]. Therefore, acquiring new knowledge and skills is necessary to implement a safe CNA, and *gamification* is a promising solution.

J. Bravo and G. Urzáiz (Eds.): UCAmI 2023, LNNS 841, pp. 205–216, 2023.
https://doi.org/10.1007/978-3-031-48590-9_20

Hands-on experiences and competition-like exercises are widely used as a learning method in cybersecurity, and Capture The Flag (CTF) contests embody the concept of gamification by rewarding and motivating the player. Even more so, Cyber Ranges represent the ideal training context by presenting more complex and elaborate environments that virtualize real-world scenarios and, as defined by NIST, "provide a *safe*, *legal* environment to gain hands-on cyber skills" [21].

Unfortunately, no virtualized cloud environment today matches the Cyber Range definition given by NIST. Current training platforms are set up using real cloud services, running up against legal limitations imposed by cloud providers (e.g., AWS penetration testing policies [4]) or against safety issues inherent to the shared nature of the cloud.

This work aims to discuss the implementation of an *isolated local environment* that is both *safe* and *legally* allows for any testing, simulating the same cloud services generally offered by the major cloud providers.

Contributions. Briefly, the main contributions of this work are:

1. The analysis of the main elements that compose a CNA and the discussion of the major security flaws they might present.
2. The implementation of a toolkit that can emulate a Cloud Provider and extend cyber range scenarios with realistic CNAs while preserving the safeness and legal requirements.
3. An assessment of the toolkit through creating a hands-on training exercise experimented in the University of Genoa's 2nd level Postgraduate Master's degree in Cybersecurity and Critical Infrastructure Protection.

Structure of the Paper. The paper is organized as follows. Section 2 briefly discusses some related work, while Sect. 3 provides the necessary technical background. Section 4 describes the architecture of the toolkit and its implementation details. Section 5 discusses a CNA's case study and its implementation, while Sect. 6 discusses the experimental setup and evaluation. Finally, Sect. 7 concludes the paper by pointing out some future directions.

2 Related Work

To the best of our knowledge, this is the first work that aims to implement a training environment for cloud security that satisfies the NIST Cyber Range definition. However, educational projects deployed on cloud provider services aim to provide participants with hands-on experience on cloud vulnerabilities. However, they do not offer a set of tools that can be utilized to deploy cloud-native applications locally. Therefore, they must leverage existing cloud providers to launch vulnerable scenarios that can be exploited to solve CTF-like challenges. Such tools can be divided into two categories, namely *Vulnerability Exploitation* and *Incident Response* projects.

Vulnerability Exploitation. Projects in this category contain a collection of vulnerable scenarios deployable on several cloud providers. Here, users acquire knowledge of common attack patterns by exploiting vulnerabilities belonging to classical categories like access control misconfiguration, privilege escalation, event injections, vulnerable CI/CD pipelines, and insecure secrets storage. Projects in this category can be very specific (i.e., focusing on some vulnerability), or - as in the case of the proposed work, may follow a more general approach. Examples of specific projects are *IAM Vulnerable* [6], *flAWS* [31], and *Sadcloud* [19]. More general projects are *CloudGoat* [28], *Serverless-Goat* [24], *Damn Vulnerable Cloud Application* [17], and *Breaking and Pwning Apps and Servers on AWS and Azure* [3].

Incident Response. Projects in this category focus on teaching the defensive mechanisms and actions that must be performed in case of threat occurrence. Therefore, they present already attacked scenarios where the user aims to track down the attacker and confine it by leveraging the tools at his disposal. Notable projects on this category are Projects such as *flAWS 2* [32], *IncidentResponseGenerator* [12], and *AWS Detonation Lab* [30].

3 Background

This section provides a high-level, cloud provider-independent introduction to Cloud Native, identifying the major components that comprise a CNA.

Cloud Native: Before Cloud Native architecture, applications were monolithic (i.e., a single and unbreakable entity, where any change could affect the whole app), with a long development process. Conversely, Cloud Native refers to a microservices architecture where each application comprises various independent functional components, that communicate through APIs and can be updated without affecting others [29].

Cloud Native Applications (CNA): CNAs rely on the Cloud Native paradigm, i.e., they are microservice-based applications running on the Cloud. Consequently, they have several advantages, like load-balancing, resilience to error and failure, cost-efficiency, automation, fast development, and easy management.

CNA Structure: Fig. 1 depicts the main components of Cloud Native Application, accessible through the mobile, web, or custom client. From a CNA management standpoint, cloud providers offer dedicated web or command-line applications to administer their components. Automation further enhances this process by providing specialized software tools and scripts that streamline tasks and workflows involved in deploying or updating applications. Notably, tools that follow the Infrastructure as Code paradigm exemplify the power of automation to manage cloud resources efficiently. Both the Client and Management application and the Automation tools interact with the CNA and its components through a unified point of access identified as the Cloud Management Services.

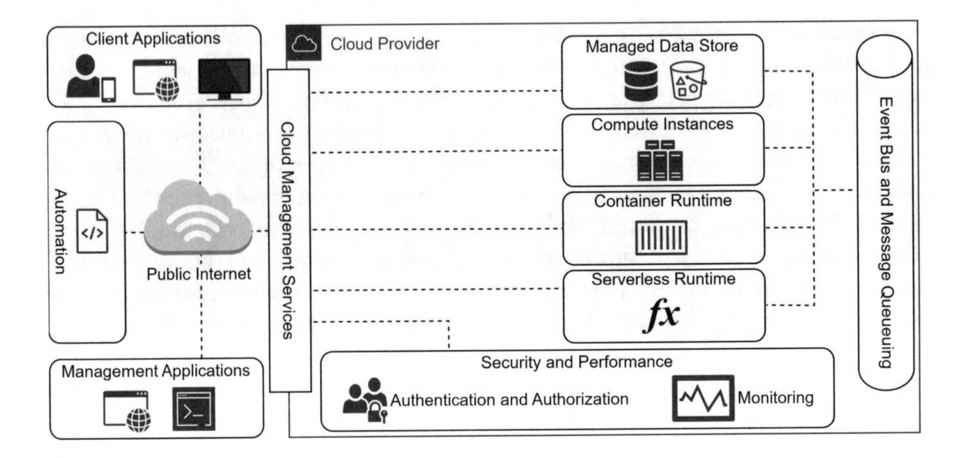

Fig. 1. Cloud Native Application components.

In the following, we provide a brief overview of the main components and their functionalities that providers can offer to develop Cloud-Native Applications (CNA):

- *Managed Data Store* refers to components that offer functionalities for data and state persistence.
- *Compute Instances* indicates the set of provided virtual machines deployable from starting images and spanning major operating systems.
- *Serverless Runtime* are event-driven services that deploy and execute program code in appropriate containers automatically, in a Function-as-a-Service fashion, without the need for the developer to manage the application runtime.
- *Container Runtime* is responsible for providing the runtime environment for microservices, providing the necessary infrastructure and resources to ensure the smooth operation of the CNA and its individual services.
- *Event Bus and Message Queuing* enables messaging and notification services to support communication between the application components.
- *Authentication and Authorization* enforces the validation of entities and their authorized actions across all services and resources, ensuring secure access and control within the provider.
- *Monitoring* collects all events and metrics belonging to the application to detect failures and anomalies and take corresponding decisions.

4 Presenting the Cyber Range Toolkit

In this section, we present our toolkit designed to create cyber range scenarios replicating the primary services offered by Amazon Web Services (AWS), a leading and long-standing cloud provider [33]. While the toolkit focuses on AWS services, it is worth noting that other cloud providers offer similar functionalities and follow similar patterns and methodologies. As a result, the outcomes gained

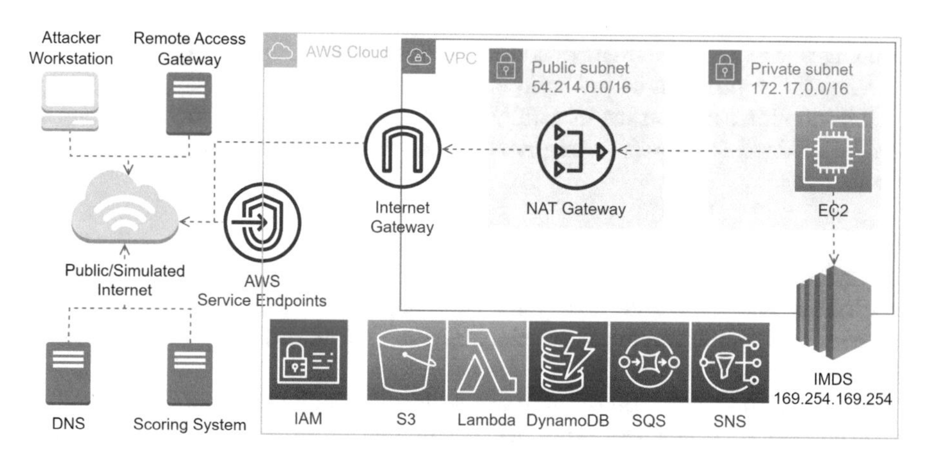

Fig. 2. Toolkit Architecture.

from using our toolkit can be regarded as applicable and generalizable across various cloud environments.

The toolkit architecture is depicted in Fig. 2.

At the core of the toolkit lies the AWS Service Endpoints component. This component embodies the Cloud Management Services of AWS, offering URL entry points that facilitate the management of services. Users can seamlessly interact with the emulated cloud instance using the official AWS Command Line Interface (CLI) and compatible infrastructure orchestrators like Terraform [14]. In its current version, the toolkit enables the instantiation of a comprehensive set of core services essential for developing cloud-native applications as introduced in Sect. 1. In particular, it supports services implementing the managed data store through object storage (S3) and NoSQL database (DynamoDB), serverless runtime (Lambda), and event bus and message queuing (SNS and SQS). Moreover, it provides a virtual networking environment (VPC) that can host compute instances (EC2) on the private IP addresses class 172.17.0.0/16 and expose the running services on the Simulated Internet using an IP address from the fake AWS public class 54.214.0.0/16. Each EC2 node can contact an emulated metadata service (IMDS) to retrieve configurations, e.g., hostname, ssh public keys, or authentication roles to access other instantiated services. Finally, the toolkit includes support for authorization and access control functionalities through the IAM component, effectively emulating AWS's Role-Based Access Control (RBAC) implementation.

A Simulated Internet connects the AWS cloud and other key components of the toolkit. In detail, a Domain Name Service (DNS) handles name resolutions and overrides the domain name assigned to AWS, pointing it toward the URL endpoint of the emulated instance. Furthermore, the Simulated Internet also hosts workstations that can be allocated to trainees acting as attackers. These workstations are available via remote desktop access provided by an authenticated Remote Access Gateway. Lastly, the Scoring System manages the challenges that trainees must tackle and offers dashboards displaying individual scores and the progress of the competition.

The toolkit can operate in a fully isolated mode, or the Simulated Internet can be bridged to the real Internet by utilizing the connectivity of the host running it.

Toolkit Implementation. The implementation follows the general architecture of scenario components of our cyber range [16,27]. In particular, containerization and Docker play a key role. Using this approach, the toolkit and the hosted CNA can coexist within a single virtual machine without imposing significant resource requirements (see Sect. 6).

For the implementation of the AWS Service Endpoints, we employ Moto [34], a Python library specifically designed for emulating programmatic interactions with AWS services during testing. Moto simplifies the process by providing fake API call responses without running the real services or offering only basic functionalities. To enhance Moto's capabilities, we adopt the *Proxy Pattern*, also known as *Monkey Patching*. This approach allows us to patch existing objects at runtime, integrating our custom code into the responses provided by Moto. As a result, we extend Moto to instantiate emulated services or implement missing functionalities. For example, we extend the EC2 API support by providing the instantiation and configuration of emulated EC2 nodes through Docker containers. We also improve the implementation of the AWS IAM component to provide better support for IAM policies and expand its capabilities to cover more emulated services.

We realize the Internet and NAT Gateway components using a containerized instance of OpenWRT [2], an open-source Linux distribution with router and firewall capabilities. When creating an EC2 instance, the Network Address Translation (NAT) and routing configurations are set accordingly through the API interface offered by OpenWRT.

The IMDS is implemented using the Python Flask [26] application framework and provides all the metadata required for configuring EC2 instances with cloud-init [7]. It supports session-oriented tokens as introduced in version 2 [5].

CordeDNS [1] runs DNS services by overriding the resolution of official AWS zones (see above) and dynamically registering names of EC2 instances.

Containers with Kali Linux [22] provide the workstations for attackers. These workstations can be accessed through a clientless remote desktop connection, made possible by a Guacamole server [35].

Lastly, the scoring system leverages CTFd [9], a mainstream open-source software that provides a framework for self-hosting CTF competitions. It offers features for managing challenges, scoring participants, and tracking progress during training sessions.

5 Case Study

The toolkit discussed in the previous section emulates cloud components that the AWS provider offers. Consequently, we apply it to a case study which is a CNA that can be hosted on the replicated AWS services.

Figure 3 depicts the architecture of the CNA which reproduces an airline web application that users can leverage to exploit the airline functionalities and, in particular, request their customer boarding pass. Concerning the latter point, the numbers in the same figure represent the steps to generate the boarding pass, namely 1) the user's email and booking number are sent to Amazon SNS, which supports a publish-subscribe model; 2) the message published by the SNS service triggers an AWS Lambda (subscriber) function; 3) the Lambda function creates the boarding pass by properly interacting with a noSQL DynamoDB service containing flight data; 4) if checks are fine, the same Lambda function generates the boarding pass and stores it on an S3 bucket; 5) the front end retrieves the boarding pass and prompts it to the user.

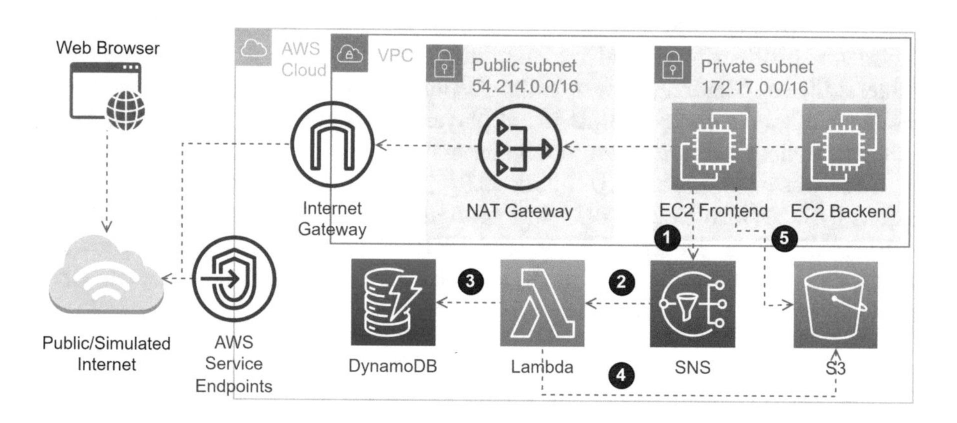

Fig. 3. Airline Web Application.

The website is hosted in a distributed fashion by leveraging AWS EC2. The first virtual machine is devoted to the website's front end, containing the website's HTML pages and all the required PHP logic needed to interact with the back end and other application components. The front end runs on an NGINX [13] server, that also supports load-balancing.

The remaining VM handles the website's back end, hosting a PosgresSQL [38] database for storing account credentials.

6 Experimental Evaluation

This section discusses how the previously introduced CNA is used to verify its educative potential, explaining how the exercise was administered and the obtained results. The goal of the implemented cyber range is to provide knowledge on the most frequent and common vulnerabilities that characterize CNAs. Therefore, the implemented case study hosted on our cyber range is a CNA vulnerable by design, presenting several flaws that malicious attackers can exploit.

To enhance the learning experience it is important to incentivize players to keep a high engagement level throughout the whole activity, and for this reason, the *gamification* [11], i.e., the use of game design elements in a non-game context, was leveraged. Gamification is especially popular in computer security, where CTF competitions have become a widely spread standard. In CTF competitions, players must discover and exploit the system's vulnerabilities to obtain the "flag", a string that acts as a reward that acknowledges the player's exploitation and effort.

Building on the positive experience [10] with Computer Science and Engineering students in both Bachelor's and Master's programs, we adopted a CTF-like approach to engage students from the 2nd level Postgraduate Master's degree in Cybersecurity and Critical Infrastructure Protection at the University of Genoa.

Fifteen students were grouped into five distinct teams for the training session. The primary objective was to equip participants with practical, hands-on experience in identifying and understanding significant and fundamental cloud vulnerabilities. Challenges were assigned points based on complexity, and each challenge offered a series of hints to aid players. Players could redeem these hints, reducing their maximum score upon successfully solving the challenge.

CTF runs on Ubuntu GNU/Linux 22.04, installed on a Virtual Machine (VM) hosted by VMWare ESXi 7.0U3 and configured with 2 Intel Xeon Gold 6252N vCPUs at 2.3GHz, 16 GB of RAM, and 40GB of disk. Each team was provided with a dedicated Kali desktop made available by the toolkit to access the CNA instance.

6.1 Notes on the CNA Vulnerabilities and the Challenges

As discussed in Sect. 2 CNAs combine classical vulnerability patterns [23] to new, cloud-specific, ones [8]. Several studies, such as the MITRE ATT&CK guidelines [37], analyze such vulnerabilities and aim to provide knowledge on some of the major techniques and tactics used by hackers in the cloud environment. We leveraged the MITRE guidelines to choose a set of vulnerabilities injected in our CNA so that attackers can exploit the tactics described in the MITRE Cloud Matrix [36], thereby obtaining a "vulnerable-by-design" CNA.

On top of the vulnerable CNA, we built different challenges that are summarized in Table 1. Here, each row presents a brief description that hints at the contents of the challenge, along with a section devoted to the core cybersecurity concepts that completing the challenge is supposed to provide. The first column shows the name of the corresponding challenge prompted to the students. Finally, the last column gives an approximate evaluation of the complexity of the whole exploitation process for each challenge.

Table 1. Challenges proposed within the vulnerable-by-design CNA.

Challenge	Description	Learining focus	Complexity
It's already 2022	Perform a SQL injection to log into the admin page	Importance of user input sanitization when used to query DBs	Beginner
Hide and S3ek	Leverage the website's *ping* tool to inject commands into the remote host	Importance of sanitizing user input to counter malicious commands injection	Beginner
The magic of EC2	Use the AWS CLI to view publicly accessible bucket contents	Misconfigurations and enforcement of appropriate access control	Beginner
Hide and S3ek	Contact the metadata service to obtain the EC2 instance user data	Knowledge about the token communication protocol used by the IMDS, along with its structure	Intermediate
The magic of EC2 II	Contact the metadata service to steal EC2 instance role credentials	Knowledge about IMDS, and EC2 roles usage	Intermediate
Thinking laterally	Gain SSH access to the second instance, by discovering a backup bucket in the first VM, containing the private SSH key	Improper practices of storing sensible credentials in insecure locations that lead to infiltration in apparently unreachable resources	Intermediate
λ	Leverage different permissions set to upload a malicious function	Lax privileges that lead to critical threats when combined	Advanced

6.2 Experimental Results

We analyzed the experimental results on the CTFs through the metrics offered by CTFd, throughout the whole competition. Due to space constraints, we just provide the collected results shown in the percentage bar graph in Fig. 4.

This view is very significant for our evaluation purpose since it allowed us to determine whether students perceived the different challenges' complexity as was intended. In this case, we were able to positively confirm that the data reflect the complexity grades associated with the different challenges; in fact, the percentages of successfully solved challenges decrease with their increasing difficulty.

Metrics show that every team has been able to exploit the mid to low-complexity vulnerabilities, meaning that they have gained a sufficient understanding of the most basic cloud vulnerabilities, which was the starting goal of the proposed exercise.

Lastly, the required setup confirms the lightweight and affordable nature of the toolkit, making it an ideal extension for cyber range scenarios with cloud services.

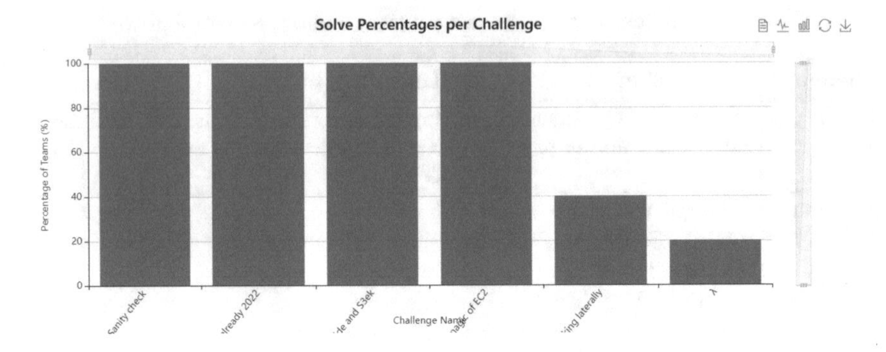

Fig. 4. Solve Percentages Graph.

7 Conclusion and Future Works

The proposed work discussed the design and implementation of a toolkit to develop a cyber range that allows hosting CNAs and performing security training based on CTF challenges. Although it may share similarities with existing training programs, this project overcomes its limitations by providing a secure and legal environment. As a matter of fact, although the implemented toolkit replicates major cloud services offered by AWS, they are not backed by a real cloud provider, meaning that it is both legal and safe to perform any kind of cyber attack on CNAs that are hosted on the proposed cyber range.

As a future extension, the toolkit should be extended and implement the remaining cloud services (e.g., monitoring and automation services) so that different use cases can be ported and various vulnerable scenarios can be enforced. Furthermore, the current EC2 service allows for the deployment of instances in containers with the same base image. Therefore, it would also be appropriate to extend this (and other) service to offer a wider range of images to choose from, along with the possibility of storing custom images in a registry. Finally, integrating a SIEM component would contribute to the realism of the proposed exercises and further enrichen the training experience once the appropriate services have also been implemented.

Acknowledgements. This work was partially funded by the NextGenerationEU project "Security and Rights in CyberSpace" (SERICS). It was carried out while Giacomo Longo was enrolled in the Italian National Doctorate on Artificial Intelligence run by the Sapienza University of Rome in collaboration with the University of Genoa.

References

1. CoreDNS: DNS and Service Discovery. https://coredns.io/. Accessed June 2023
2. Openwrt. https://openwrt.org/. Accessed Sept 2022
3. Appsecco: Breaking and Pwning Apps and Servers on AWS and Azure - Free Training Courseware and Labs. https://github.com/appsecco/breaking-and-pwning-apps-and-servers-aws-azure-training. Accessed Sept 2022

4. AWS (Amazon Web Services): Penetration Testing. https://aws.amazon.com/security/penetration-testing/. Accessed Aug 2022
5. AWS (Amazon Web Services): Use IMDSv2. https://docs.aws.amazon.com/AWSEC2/latest/UserGuide/configuring-instance-metadata-service.html. Accessed June 2023
6. Bishop Fox: IAM Vulnerable. https://github.com/BishopFox/iam-vulnerable. Accessed Aug 2022
7. Canonical: cloud-inig the standard for customising cloud instances. https://cloud-init.io/. Accessed June 2023
8. Costas, L., Sokratis, K.K., Pitropakis, N.: Cloud security, privacy and trust baselines. In: Vacca, J.R. (ed.) Cloud Computing Security, chap. 4. CRC Press (2016)
9. CTFd: CTFd. https://github.com/CTFd/CTFd. Accessed Sept 2022
10. Demetrio, L., Lagorio, G., Ribaudo, M., Russo, E., Valenza, A.: ZenHackAdemy: ethical hacking @ DIBRIS. In: Proceedings of the 11th International Conference on Computer Supported Education. SCITEPRESS - Science and Technology Publications (2019). https://doi.org/10.5220/0007747104050413
11. Deterding, S., Dixon, D., Khaled, R., Nacke, L.: From game design elements to gamefulness: defining "gamification". In: Proceedings of the 15th International Academic MindTrek Conference: Envisioning Future Media Environments, pp. 9–15 (2011)
12. DisruptOps: IncidentResponseGenerator. https://github.com/disruptops/IncidentResponseGenerator. Accessed Sept 2022
13. F5: NGINX: Advanced Load Balancer, Web Server, and Reverse Proxy. https://www.nginx.com/. Accessed June 2023
14. HashiCorp: Terraform Documentation. https://www.terraform.io/docs. Accessed Sept 2022
15. Kratzke, N., Quint, P.: Understanding cloud-native applications after 10 years of cloud computing - a systematic mapping study. J. Syst. Softw. **126**, 1–16 (2017)
16. Longo, G., Orlich, A., Musante, S., Merlo, A., Russo, E.: MaCySTe: a virtual testbed for maritime cybersecurity. SoftwareX **23**, 101426 (2023). https://doi.org/10.1016/j.softx.2023.101426
17. Maxime Leblanc: Damn Vulnerable Cloud Application. https://github.com/m6a-UdS/dvca. Accessed Sept 2022
18. Ministry of Innovation Technology and Digitalisation: Strategia Cloud Italia (2021). https://docs.italia.it/italia/cloud-italia/strategia-cloud-italia-docs/it/stabile/index.html. Accessed Aug 2022
19. NCC Group: Sadcloud. https://github.com/nccgroup/sadcloud. Accessed Sept 2022
20. Netflix: Four Reasons We Choose Amazon's Cloud as Our Computing Platform. https://netflixtechblog.com/four-reasons-we-choose-amazons-cloud-as-our-computing-platform-4aceb692afec. Accessed Sept 2022
21. NIST (National Institute of Standards and Technology): Cyber Ranges. https://www.nist.gov/system/files/documents/2018/02/13/cyber_ranges.pdf. Accessed Aug 2022
22. OffSec Services: The most advanced Penetration Testing Distribution. https://www.kali.org/. Accessed June 2023
23. O'Hara, B.M.: CCSP (ISC)2 Certified Cloud Security Professional Official Study Guide. Sybex (2017)
24. OWASP: Serverless-Goat. https://github.com/OWASP/Serverless-Goat. Accessed Aug 2022

25. OWASP (Open Web Application Security Project) Foundation: OWASP Cloud-Native Application Security Top 10. https://owasp.org/www-project-cloud-native-application-security-top-10/
26. Pallets: Flask, web development, one drop at a time (2010). https://flask.palletsprojects.com/en/2.2.x/. Accessed Sept 2022
27. Raimondi, M., Longo, G., Merlo, A., Armando, A., Russo, E.: Training the maritime security operations centre teams. In: 2022 IEEE International Conference on Cyber Security and Resilience (CSR), pp. 388–393 (2022). https://doi.org/10.1109/CSR54599.2022.9850324
28. Rhino Security Labs: CloudGoat. https://github.com/RhinoSecurityLabs/cloudgoat. Accessed Aug 2022
29. Rob, V., Stive, S.: Architecting Cloud Native .NET Applications for Azure. Microsoft Developer Division, .NET, and Visual Studio Product Teams (2020)
30. Nolette, R.: AWS Detonation Lab. https://github.com/sonofagl1tch/AWSDetonationLab. Accessed Sept 2022
31. Scott Piper. Summit Route: flAWS. http://flaws.cloud/. Accessed Sept 2022
32. Piper, S.: Summit Route: flAWS 2. http://flaws2.cloud/. Accessed Sept 2022
33. Services, A.W.: AWS Named as a Leader in the 2022 Gartner Cloud Infrastructure and Platform Services (CIPS) Magic Quadrant for the 12th Consecutive Year. https://aws.amazon.com/it/blogs/aws/aws-named-as-a-leader-in-the-2022-gartner-cloud-infrastructure-platform-services-cips-magic-quadrant-for-the-12th-consecutive-year/. Accessed Sept 2022
34. Pulec, S.: Moto - Mock AWS Services. https://github.com/spulec/moto. Accessed Sept 2022
35. The Apache Software Foundation: Apache Guacamole. https://guacamole.apache.org/. Accessed June 2023
36. The MITRE Corporation: Cloud Matrix. https://attack.mitre.org/matrices/enterprise/cloud/. Accessed Sept 2022
37. The MITRE Corporation: MITRE ATT&CK. https://attack.mitre.org/
38. The PostgreSQL Global Development Group: PostgreSQL: The World's Most Advanced Open Source Relational Database. https://www.postgresql.org/. Accessed June 2023

Monitoring Energy Consumption via Middleware Services

Carlo Ferrari(✉)

Department of Information Engineering, The University of Padova, Padova, Italy
carlo.ferrari@unipd.it
http://www.dei.unipd.it

Abstract. A very basic attack to computing devices involves wasting their energy in useless computation. The countermeasures have to defer as much as possible the shutdown time due to a definite lack of energy anytime it is completely unfeasible to restore energy by plugging the device at some recharging station. This paper propose to add, at the midlleware level, an "Energy Server (ES)" that monitors the energy consumption from all running components and applications within a single devices, and it is able to discover any potential situation of energy wasting. The countermeasures ES can apply range from underscheduling those components that can use the most part of the available energy to slowing down the whole processor(s) up to asking the human intervention in the most critical situation. The ES, as an infrastructural element at the middleware level, can be modeled as an event-based systems receiving notifications through a shared-memory approach (instead of using messages). ES directly cooperates with the OS keeping a list of running and frozen local services and processes either specific or shared among applications. The decision process stands on collected data (and histories) of standard usages from the most common services in order to discover both applications energy eaters components and overuses of some specific "chunks of software".

Keywords: Energy consumption evaluation · Middleware for Monitoring Services · Energy-aware systems

1 Introduction

Big data processing, number crunching, massive communications among components in distributed environments are examples of computing activities that eat energy at any level and the energy consumption is higher when results are expected in shorter time. Currently the most common architectures are parallel ones at the inner level of CPU and GPU and distributed ones at the outer level ranging from classical multitiered client-server to most modern edge-based arrangements and organization that better match the need of distributed intelligence for pervasive and ubiquitous contexts.

J. Bravo and G. Urzáiz (Eds.): UCAmI 2023, LNNS 841, pp. 217–222, 2023.
https://doi.org/10.1007/978-3-031-48590-9_21

Short, medium and long range mobility adds complexity due to the request of reconfiguration of those directly interacting devices while keeping alive the cooperation and sharing of different components at the application levels. Mobility opens to different scenarios in term of access to energy sources that can be rougly divided in two main categories: autonomous producer of energy (like cars with solar cells) and devices equipped with energy storage either with or without recharging capabilities via direct interacion with a docking station.

Complex classical systems involve simple basic IoT devices that may directly benefit from solar energy together with far backends in specific location (like those in server farms) that optimize efficient power supply and conditioning. This description represents widespread client-server structures with thin clients for interfacing with users and well-designed server for processing and long term memory and it is reasonable if we can choose to concetrate and centralize the "processing intelligence" in specific location. The actual scenario for a ubiquitous world, perceived as a distributed computational intelligence, is not supported by the technical paradigm of a strict separation among tiers (then deployed on different devices over a net) but instead ask for more (pre)computational activities that abstract concept from (raw)data generally obtained from the environment: in other word IoT devices at the periphery of the net are supposed to be in charge of more complex computational task. As computing use energy and fast computing produces heat (cooling systems use energy as well) energy consuption at the perifery of the net is one of the key issues in modern systems.

The attention of the scientific community towards Energy Aware Systems started more than a decade ago mainly driven by the introduction of small and tiny devices usually acting as components over a sensors net and the main focus was on saving and restoring energy in order not to stop the data acquisition and data forwarding processes [1,2]. Energy-aware hardware design and energy storage were the main issues followed at a later time by security related issues that called for newer communication protocols [3,4]. While mobility has been (and still is) one of the driving force of research on energy-aware systems, another issue now emerging is related to scalability. Actual systems are not made anymore of homogenous components located in a small area (maybe in large number) but they are formed by heterogeneous devices some of them located at the edge of the net interacting with far backend [5]. Moreover the weight of (pre)computation is becoming higher at the edge in order not to load net services and backend servers. Then scalability arises due to the increasing and time variant number of heterogenous systems that could either share or compete for energy sources at the same physical location. The situation is even worse when considering the modern evolution of DoS attacks: its basic form involves using IoT devices energy in useless computation, like answering at fake requests. The result is wasting and soon exhausting energy for the most part of computing elements in a single system: the effect is worse that starvation as the system soon can collapse. It is obvious there is no general solution to these kind of attacks and the correct approach is considering the choice of different fault-tolerant strategies according to the type of requested process, the time requirements for producing results and some hardware set-up for energy restoration. Prediction about either common or standard usage must be exploited as well. The key point is to defer as much as possible the shutdown time due to the lack of energy when it is unfeasible to restore energy by plugging the device at some recharging station. The proposed paper requires to add a kind of "Energy Server (ES)" that monitor the energy consumption from all running components and applications within a single devices, in order to discover any potential situation of energy wasting. The countermeasures ES can applies ranges from underscheduling those components that can use the most part of the available energy to slowing down the whole processor(s) up to asking the human intervention in the most critical situation. ES directly cooperates with the OS keeping a list of running and frozen local services and processes either specific or shared among applications. We would address the basic design principles that can be implemented in the middleware (becoming a true new middleware added service) for the ES. The role of the ES is manyfold: at least it is required to discover and highlight all situations where there is a significant waste of energy. A proficient ES is supposed to be able to delay as much as possible the shortage of energy by suggesting to the OS proper changes to the scheduling procedures. Moreover as a further step, ES have to suggest and activate some procedure for restoring on-board energy (whenever it is possible) while considering a gracefully and secure shutdown as a last opportunity. Finally an ES should be able to manage log information about the processing system it is tied to, to accumulete experience about the use and role of each specific component. In the next paragraph we will describe the basic middleware primitives, services and functionality of the ES while in the following paragraph we will summarize the main design principle and characteristics of the overall organization of the ES.

2 Middleware Services for Energy Management

According to the idea that middleware is a software stratum between platforms (i.e. hardware, OS and communication protocols) and applications (i.e. user processes) it is possible to design some primitive for the ES that abstract basic services. Starting from the consideration that not all processes/applications have to be monitored nad observations can concentrate to the most critical ones, the ES should admit some registration/deregistration primitive in order to explicitly let applications put under attention specific related processes. Registration can be either requested or forced by the platform itself anytime new processes are going to be started: the registration is in the push–mode form. At the same time trusted application can ask directly to be monitored and require the platform to be registered directly (pull-mode registration). Deregistration takes places anytime the platform realizes that some processes/components are not worth to be monitored anymore (push–mode deregistration) while deregistration requests from the application are not directly accepted as they have to be authorized by the platform. Registration/deregistration operations act on a table: each entry is associated to a single process to be monitored and has some additional values: at present, we consider to add in/out time, either soft or hard processing time limit (if applicable), qualitative evalutation related to energy consuption per process (like low, medium, high), qualitative recommandations for energy attribution and so on. In order to keep the table under reasonable dimensions, each entry has a time-limit: when it expires the entry is moved to a another table of expired entries (saved on secondary memory). Both the registration and deregistration primitives can accept some parameters (time, delay, and so on), that specialize each entry: each parameter admits a default value in order to avoid empty values in the table. Registration/deregistration operations are not related to process scheduling in order not to waste time to follow the high frequency associaton of processors cycle to the processes themselves. Instead they are related to the true activity of process to be monitored. It is worth to note that is seems reasonable some entry are considerent permanent resident in the table: these are tied to process alive as long the device is alive.

Information about energy usage by processes are obtained using some query primitives (called metering primitives) that get information either about the use of resources by each registered process, or about the status of the energy resources as well. As it is not possible to exactly measure specific energy consumption, those primitive compute an estimate considering the computing time associated to the process plus the clock frequency. An overhead due to the activity of the OS related to scheduling and context switching is added. Queries about the battery status is easier because any OS has primitives for evaluating the battery charge.

The ES has to decide about the activation of countermeasures. Decisions are taken comparing those figures obtained from the metering primitives with some reference values. These values can be either absolute and fixed at bootstrap or variable according to the true state of the device. In the latter situation we consider adaptive reference values that can be modified (either raised or lowered) by proper ES primitives, considering the actual trend of processing and some information about near-to-be-scheduled processes.

Countermeasures directly act on scheduling parameters mainly through a delay primitive, that suggest new precedence parameters of processes to the OS. The effect of lowering the precedence is a delay in execution: the precedence of delayed processes will never be restored to its original value. A second, stronger primitive, act on clock frequency. Slowing down clock can help in saving energy but its effect is on all the set of ready processes. This action admit a counter action to restore the clock frequency ot its original value. Finally it is worth to consider that high level energy consuming processes can be temporarilly frozen whenever their execution could exhaust the battery in a very short period: the ES has the couple of freeze/resume primitive that inform the OS about those processes whose state have to be saved in secondary memory for being resume at a later time.

The last type of primitives of the middleware are for logging and reporting purposes. Writing proper information on file help system administrators to analyze the ES behavior for discovering weakness in the decision strategies and any fault of the systems itself.

The ES has a clearly local scope but it is reasonable to identify some functions that interact with other ES at other node belonging to the same system. Interaction is useful for sharing parameters values, organizing the access to recharging station and checking the opportunity to move process from one node to another node (better shaped about energy). These primitives have to use secure connections beteween nodes and their functionality is mirrored between the node involved in the communications.

3 Main Features of the ES

Monitoring activities may activate proper procedures both to keep the system in a kind of healty state (wrt energy availability) and to proper react to unexpected situations. The advantages and drawbacks of syncronous and asyncronous monitoring are well know in the literature and technical history of Computer Engineering and the choice between them depends on specific factors strictly related to the applicative scenario. The ES is a middleware component mainly operating at the local level, i.e. at any single computing device, supervising energy for processing and, as a consequence, it has to consume as less as possible energy for its own processing. Hence the basic mechanism to be implemented for the ES is asyncronuos monitoring: the ES is quiescet unless the values of alert/alarm variables overcome established default threshold. The ES, as an infrastructural element at the middleware level, can be seen as an event-based systems: observers processes are directly tied to alert/alarm variables and ES's subscription to event notification are set up during the initialization phase, in a sort of hard coded fashion. Due to the local nature of the ES, notifications (form the observers to the ES) are implemented exploiting a shared-memory approach, instead of using messages, saving the cost for message formation and forwarding. Metering primitives access the "notification pool" to deposit new measures: decision queries are activated any time values in the pool exceed thresholds and generate data for decision directly read by the ES. The "notification pool" is managed according the reader-writer paradigm, using the basic mechanism for software interruptions of the local OS.

The decision process stands on either directly collected data or undirectly on compound figures from some data fusion activity. All the process is strongly dependent on threshold and their default values that may somhow vary during the ES activity. Threshold initial set up may benefit from collected data (and histories) of standard usages from the most common services in order to discover both applications energy eaters components and overuses of some specific "chunks of software". These data are managed by logging and reporting primitives. Saving usage statistics on secondary memory requires energy and it may happen that under some circumstaces these activity have to be suspended and part of the usage data are lost. Off line maintenance can help in reducing the effect of these kind of faults, by re-setting proper default values for the decision variables.

4 Conclusions

In this paper we presented the design and main features of a specific new service at the middleware level (the ES–"Energy Server") in charge of monitoring and managing energy consumption for edge-devices in distributed environments. Those devices are not simple IoT ones but are in charge of some part of the overall processing in order not to overload the backend servers and the net. The ES is an event-based system and notifications are via a shared pool of monitoring variables. The ES interfaces to the OS and platforms with some primitives both for reading the system status (processes resource usage and precedences) and for modifying scheduling criteria in order to slow down or stop energy eater processes. More sophisticated mechanisms can be added in order to set up and adapt threshold values for decision: tracing, histories and log files can help in forecasting potentially dangerous situations bringing to sudden shutdown of devices. The systems is currently under development and experiments on its effectiveness are following.

Acknowledgements. CF has been partially supported by the University of Padua project DOR2023.

References

1. Kanoun, O., et al.: Energy-aware system design for autonomous wireless sensor nodes: a comprehensive review. Sensors **21**, 548 (2021). https://doi.org/10.3390/s21020548
2. Kyung, C., Yoo, S.: Energy-Aware System Design: Algorithms and Architectures. Springer (2011). https://doi.org/10.1007/978-94-007-1679-7
3. Hoffmann, J., Kuschnerus, D., Jones, T., Hubner, M.: Towards a safety and energy aware protocol for wireless communication. In: 13th International Symposium on Reconfigurable Communication-Centric Systems-on-Chip (ReCoSoC), Lille, vol. 2018, pp. 1–6 (2018). https://doi.org/10.1109/ReCoSoC.2018.8449380
4. Conti, V., Ziggiotto, A., Migliardi, M., Vitabile, S.: Bio-inspired security analysis for IoT scenarios. Int. J. Embed. Syst. **13**(2), 221–235 (2020). https://doi.org/10.1504/IJES.2020.108871
5. Yang, K., Sun, P., Lin, J., Boukerche, A., Song, L.: A novel distributed task scheduling framework for supporting vehicular edge intelligence. In: 2022 IEEE 42nd International Conference on Distributed Computing Systems (ICDCS), Bologna, pp. 972–982 (2022). https://doi.org/10.1109/ICDCS54860.2022.00098.

FP-H: A Real-Time Energy Aware Scheduler with Fixed Priority Assignment for Sustainable Wireless Devices

Maryline Chetto(✉)

Nantes Université, École Centrale Nantes, 44000 Nantes, France
maryline.chetto@ls2n.fr
https://www.ls2n.fr/

Abstract. This paper addresses the scheduling issue in a real-time computing system such as a wireless sensor node which is supplied with regenerative energy present in the environment. We consider preemptive task scheduling with fixed priority assignment. We propose a novel energy harvesting aware scheduling approach, namely FP-H. We show how processing time and energy should be assigned to the deadline constrained tasks in a short-term perspective so as to guarantee energy neutrality whenever possible.

Keywords: Autonomous sensor node · Fixed Priority Scheduling · energy harvesting

1 Introduction

Advancements in energy, computing and wireless communication technology allow to offer new services for people and make smart environments and better quality of life. For example, traffic monitoring, car parking monitoring, space detection and public transportation monitoring are use cases for smart cities. Batteries are the obvious way of powering wireless devices. However, regular battery replacement or recharging is vital to ensure longtime operation for them. Such a requirement generally implies high maintenance costs, especially when the devices are accessible with difficulty. Note in addition that millions of sensors may be deployed around the city. The sensors have to work autonomously for very long periods of time without any supervision or human intervention. Smart city applications will expand only with a reduced operational cost and increased sustainability model for these low powered and low maintenance sensors. As battery powered sensors stop working after several years, replacing millions of primary batteries leads to environmental pollution. Energy Harvesting (EH) then appears as a potential alternative to address this autonomy issue (see Fig. 1). Ambient light as mechanical energy can be drawn from the environment so as to supply small electronic devices including wireless sensors quasi perpetually [1].

Supported by Nantes Université.

J. Bravo and G. Urzáiz (Eds.): UCAmI 2023, LNNS 841, pp. 223–234, 2023.
https://doi.org/10.1007/978-3-031-48590-9_22

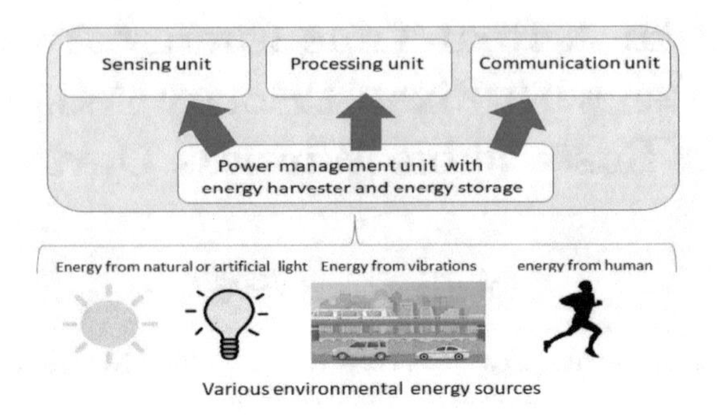

Fig. 1. Energy harvesting to supply small devices

Nonetheless, relying on variable environmental energy in a sensor node makes it challenging to guarantee reliable operation over the lifetime horizon. Thus, new methods and techniques are necessary to assess the longtime behavior of any EH-powered device which, in addition may have real-time constraints. The common characteristic of the so-called RTEH (Real-Time Energy Harvesting) systems is periodicity of activities that generally involve sampling, processing the sensed value, transmitting data, etc. [2,3] as depicted in Fig. 2. Energy neutrality is the property of an RTEH system that first should consume no more energy than the harvested energy and second should respect its timing requirements in every circumstance. To make a system energy neutral will require to identify the available power output of the harvester such as solar panel, the capacity of the energy storage unit and the energy which is consumed by the different tasks in operation.

Clearly, the challenging questions to be solved in such a system are real-time scheduling [4,5], power management and dimensioning. Firstly, how to assign a priority to each task in accordance to its importance and/or urgency? Secondly, how to dynamically adapt the activity of the processing unit so as to subsist perpetually, given the profile of the energy source and the timeliness requirements of the tasks? Thirdly, how to define the size of both the energy storage unit and the harvester in order to guarantee an acceptable quality of service?

The rest of the sections in the paper is organized as follows. In Sect. 2, a qualitative review on scheduling techniques for RTEH systems is carried out. In Sect. 3, the details of the system model under consideration, for next-generation autonomous real-time systems, are presented. Section 4 describes a new energy harvesting aware scheduler, named FP-H. A discussion on applicability of the scheduler is presented in Sect. 5 followed by the conclusion in Sect. 6.

2 Related Works

In this section, we survey the previous studies which are directly related to the proposed approach. The research on the design of self-sustainable devices started at the beginning of the 2000s. The classical real-time schedulers including RM (Rate Monotonic) and EDF (Earliest Deadline First) [6] fail in energy harvesting systems where the supply energy is intermittent. In the latters, energy must be treated as an equally important resource as time. In particular, we have to characterize the tasks by both processing time and energy consumption. And we have to achieve the online monitoring of available energy in the storage unit as well as the online prediction of environmental energy produced in near future.

The first work in the literature that explored task scheduling in monoprocessor RTEH systems is reported in [7]. It addressed frame-based tasks with voltage and frequency scaling capabilities. The Lazy Scheduling Algorithm (LSA) proposed by Moser et al. [8] is an optimal EDF-based algorithm based on as late as possible policy which applies to any set of deadline constrained tasks. Liu et al. extended LSA with EA-DVFS and HA-DVFS which are DVFS-based algorithms [9,10]. They slow down the processor so as to save energy and speed up task execution in case of overflowing harvested energy. In [11], an optimal preemptive fixed-priority scheduling algorithm called PFP_{ASAP} was proposed, assuming a constant power source.

In [12], EDF was proved to be the best non idling scheduler. In [13], Chetto presents a strongly optimal scheduler, namely ED-H, for the general RTEH model with no restriction on task arrival profile and energy source profile. ED-H is an idling variant of EDF where two key values are computed on-line: slack time and slack energy. The dynamic power management joined to the EDF priority assignment rule guarantees optimality in terms of scheduling and energy neutrality whenever possible. An exact schedulability test is given for a generic set of deadline constrained jobs, assuming accurate prediction of the incoming energy budget. In summary, most research works addressed the EDF dynamic priority assignment rule [14]. Even if EDF allows higher processor utilization than fixed-priority schedulers, this scheduler is not commonly integrated in the commercial RTOS.

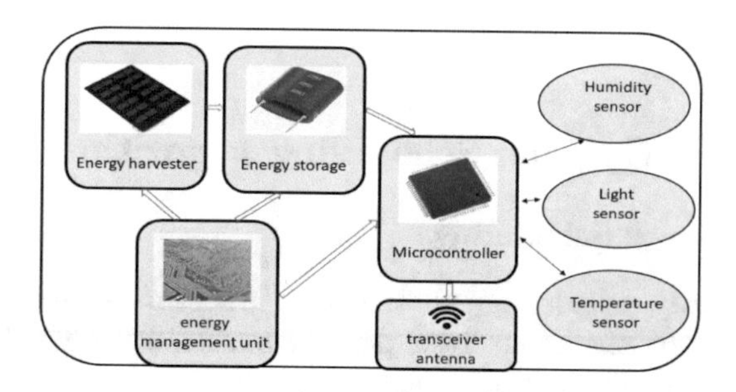

Fig. 2. Framework of an autonomous sensor node

2.1 System Model and Assumptions

Hereafter, we address an RTEH system that consists of three major parts: energy harvester, computing module supporting the real-time software and rechargeable energy storage with limited capacity.

2.2 System Model

We consider a real-time embedded system which is supplied from an energy source through an energy harvester such as solar panel. The energy harvested from time t_1 to t_2 is calculated with following formula $E_s(t_1, t_2) = \int_{t_1}^{t_2} P_p(t)dt$ where $P_p(t)$ is the worst case charging rate (WCCR).

The energy storage such as a rechargeable battery allows to continue operation even when no energy is harvested from the environment. It has nominal capacity C. It stores the extra amount of energy harvested for immediate or future use. We assume as negligible energy wasted in charging and discharging the battery.

At a given time t we have $C_{min} \leq C(t) \leq C_{max}$ where $C(t)$ is the amount of energy available in the storage at time t.

The real-time software that we are interested has independent jobs which may be the invocation requests of N periodic tasks. $\tau = \{\tau_1, \tau_2, ..., \tau_N\}$. If τ_i is a periodic task, the jobs generated by τ_i along time are also statically specified. We will focus particularly on the jobset generated by τ because the source energy is variable and the scheduling sequence varies according to different time operational conditions.

We will consider a generic set of jobs $J = \{J_1, J_2, ..., J_n\}$. J_i is completely specified by four-tuple (r_i, C_i, E_i, d_i). It respectively gives the release time, worst case execution time (expressed in time units and normalized to processor computing capacity), worst case energy consumption (expressed in energy units such as joule) and deadline of J_i. Energy consumed by the processor for executing any job is not necessarily proportional to the computation time of that job. J_i has to receive C_i units of execution and E_i units of energy in the interval $[r_i, d_i)$. $d_{Max} = \max_{0 \leq i \leq n} d_i$ and D is the longest relative job deadline i.e. $D = \max_{1 \leq i \leq n} (d_i - r_i)$. Energy consumed by the jobs on $[t_1, t_2)$ is denoted $E_c(t_1, t_2)$.

3 FP-H: The Optimal Scheduling Algorithm

3.1 Fixed Priority Scheduling

Under fixed priority scheduling, all the jobs which are generated by a given periodic task inherit the same fixed priority statically assigned to that task. The jobs then compete for the processor at run time using their priority. In this paper, we are concerned with fixed task priority assignment or fixed job priority assignment. The scheduling issue in RTEH systems is twofold: first how to assign priorities to jobs and second how to decide when to execute a job and when to let the processor in the sleep mode. Consequently, priority assignment issue and processor management issue have to be considered independently to design energy harvesting aware scheduling algorithms.

Hereafter, we consider the following definition of optimality: Suppose that a priority assignment is given, say FP. A scheduler will be said optimal if it finds a valid schedule (i.e. with no deadline missing) for every FP-schedulable jobset (i.e. at least one valid schedule exists for this jobset with the priority assignment FP).

3.2 Clairvoyance and Idling Requirements

Energy limitation and variation may affect respect of timing requirements and schedulability of jobs. Figure 3 enables us to show that deadline missing can happen if a scheduling scheme does not consider future production and future consumption of energy to take its online decision. In other words, an efficient scheduling algorithm should be clairvoyant.

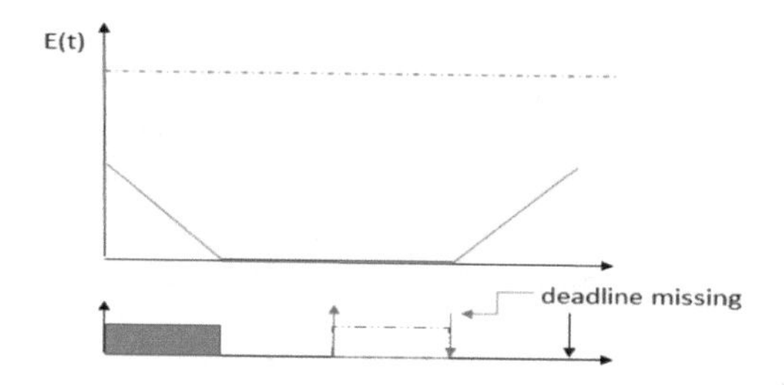

Fig. 3. Deadline missing because of no clairvoyance.

For example, Fig. 3 depicts a job which executes and consequently consumes energy. Thereafter, another job releases and no sufficient energy is available for it so as to complete its execution by its deadline whatever its priority. Considering future arrivals of jobs as well as their energy requirements is required to decide whether to put the processor in the active mode or in the sleep mode. The schedule depicted in Fig. 4 illustrates that We have to examine not only the energy that is currently stored in the SM but also the energy that will be produced by the environmental source and consumed by jobs in future.

3.3 Central Definitions

Consider a FP-H schedulable jobset at the current time t_c. We assume that J_c is the highest priority job ready for execution at t_c. We address the question of deciding if J_c can be executed while avoiding a future deadline missing caused by energy starvation. We have shown that this requires computation of a variable called *preemption slack energy* at t_c. If the preemption slack energy is zero,

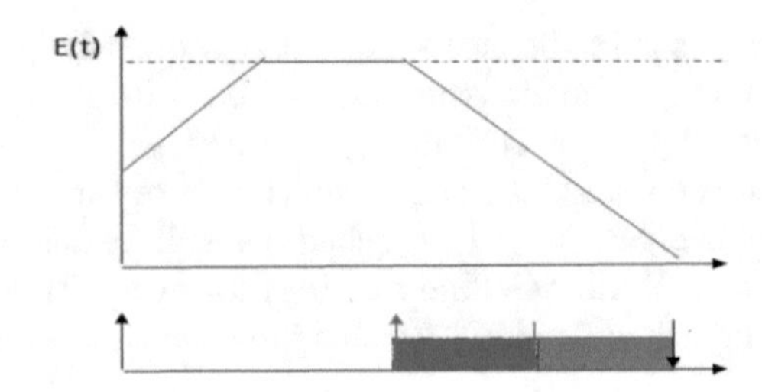

Fig. 4. Smart power management with no deadline missing.

the processor has to be put in the sleep mode for battery recharging and the duration will be equal at most to the slack time computed at t_c. Let us recall that the slack time is the maximal time interval for the processor to stay idle while guaranteeing no deadline missing.

Let us show how to identify the slack time at current time t_c for a generic PA-schedulable jobset which is not necessarily issued from periodic tasks. For clarity, let us recall useful definitions.

- A priority level-j busy period is a time interval during which jobs, of priority π_j or higher, that were released at or after the start of the busy period, but before its end, are either executing or ready to execute.
- Assume that for each priority level-j, there is only one job, say J_j. Let α_j be the set of time instants called scheduling points of job J_j defined as:

$$\alpha_j = \{t;\, t = r_k,\, \pi_k > \pi_j,\, r_j < r_k < d_j\} \cup \{d_j\} \tag{1}$$

- The slack time of job J_j at time t_c, denoted by $ST_j(t_c)$ is the largest duration such that, if jobs with a priority higher than J_j start within a delay at most equal to $ST_j(t_c)$, J_j will meet its deadline.
- The slack time of jobset J at time t_c, denoted by $ST(t_c)$ is the largest duration such that, if the execution of any job starts with a delay at most equal to $ST(t_c)$, all jobs will meet their deadline.

It was proved in [15] how to compute the slack time of job J_j at time t_c and finally the slack time of jobset J at time t_c.

Proposition 1.

$$ST_j(t_c) = \max_{\substack{t \in \alpha_j \\ t > t_c}} (t - t_c - \sum_{\substack{\pi_k \geq \pi_j \\ t_c < r_k < t,}} C_k) \tag{2}$$

Proposition 2.

$$ST(t_c) = \min_{t_c \leq r_j < d_c} ST_j(t_c) \tag{3}$$

From Proposition 1, it clearly results that if there is some time t_c and some job J_j such that $ST_j(t_c) = 0$, the processor has to be busy at t_c in order to guarantee the time validity of the FP-H schedule. To guarantee the respect of all job deadlines, we need to compute the level-j slack time for all priority levels and take the smallest one as the slack time of the jobset, as stated in Proposition 2.

In a similar way to the time domain, we introduce the notion of slack energy for the energy domain:

- The slack energy of job J_j at time t_c, denoted by $SE_j(t_c)$ is the largest amount of energy which may be consumed from t_c by jobs with a lower priority that guarantees no deadline missing for J_j caused by energy starvation.
- The preemption slack energy of the jobset J at current time t_c is the largest amount of energy which may be consumed by the currently active job J_c that guarantees no deadline missing caused by energy starvation for future higher priority jobs.

Proposition 3.

$$SE_j(t_c) = \max_{\substack{t \in \alpha_j \\ t > t_c}} (E(t_c) + E_p(t_c, t) - \sum_{\substack{\pi_k \geq \pi_j \\ t_c < r_k < t,}} E_k) \tag{4}$$

Proposition 4.

$$PSE(t_c) = \min_{\substack{\pi_j > \pi_c \\ t_c < d_j < d_c}} SE_j(t_c) \tag{5}$$

If there is some time t_c and some job J_j such that $SE_j(t_c) = 0$, Proposition 3 says that the processor has to either be idle or execute a higher priority job at t_c in order to guarantee the energy validity of the FP-H schedule i.e. absence of deadline missing caused by energy starvation. Consequently, if there is some time t_c such that $PSE(t_c) = 0$, the processor has to be idle from time t_c.

3.4 Informal Description

As the classical fixed priority scheduler FP, the energy aware scheduler FP-H still preemptively schedules the jobs according to a given priority assignment rule FP. Before authorizing the highest priority job to be executed by the processor, the residual energy in the battery must be sufficient to supply it. Furthermore, the energy consumed by it must not provoke energy starvation for future higher priority jobs. In addition to the interference with other jobs with higher priorities a job may be postponed because of necessary processor idling. This delay occurs because the processor cannot continue working without injuring energy starvation.

The online problems the power management procedure has to deal with are first to decide whether the processor can enter the active mode and if so, second to compute the maximum amount of energy that may be consumed for preserving the energy feasibility of all the subsequent higher priority jobs. Let us note that, by definition of priority, J_c cannot have any impact on the execution of any job with priority less than π_c. Thus, let us introduce the so-called *preemption slack energy of the jobset J at time t_c* denoted by $PSE(t_c)$. An idle time is forced if $PSE(t_c) = 0$. Otherwise, the processor is authorized to be busy consuming at most $PSE(t_c)$ units of energy until J_c be finished or preempted by a higher priority job.

3.5 The Scheduling Scheme

Hereafter, $L_r(t_c)$ is the list of uncompleted jobs in J ready for execution at t_c. The FP-H scheduling algorithm uses the following rules:

- At any time t_c, the priority order defined by the FP priority assignment rule allows to select the future running job in $L_r(t_c)$.
- The processor is idle in $[t_c, t_c + 1)$ if one of the following conditions is satisfied:
  1. $L_r(t_c) = \emptyset$.
  2. $L_r(t_c) \neq \emptyset$ and $E(t_c) = 0$.
  3. $L_r(t_c) \neq \emptyset$ and $PSE(t_c) = 0$
- The processor is busy in $[t_c, t_c + 1)$ if one of the following conditions is satisfied:
  1. $ST(t_c) = 0$.
  2. $E(t_c) = C$.
- The processor can be busy or idle in $[t_c, t_c + 1)$ otherwise.

These rules say that the processor should be necessarily idle if the energy storage is deplenished or if execution of any job will imply energy starvation for at least one future occurring job (i.e. the system has no preemption slack energy). Recharging power is wasted only when there are no ready jobs and the storage is full at the same time. Decisions of FP-H are based on computation of $PSE(t_c)$. This supposes to know short term prediction of the energy harvesting rate . Methods have been developed which aim at providing prediction with high accuracy, low computation complexity and low memory requirement [16]. The energy storage stops to recharge if the slack time is zero. Methods for computation of slack time are reported in [4]. FP-H assumes here that the energy thresholds are respectively 0% and 100% of the storage capacity. Nevertheless, other threshold may be specified, thus shortening duration of the discharging and recharging phases and increasing the number of switches between active mode and power-down mode of the processor.

3.6 Optimality Statement

Theorem 1. *The scheduling algorithm FP-H is optimal for any priority assignment rule.*

Theorem 1 was established in [15]. It says the following: if FP-H does not success in building a valid schedule for any FP-schedulable jobset J, then no other processor management policy that respects the priority assignment FP will be able to build a valid schedule.

4 Illustrative Example

The following is an example to illustrate the energy aware scheduler FP-H and the performance gain compared to the classical scheduler FP. We consider an RTEH system composed of the jobset J whose time and energy parameters are in Table 1. The hardware platform has an energy storage unit with capacity $C = 10$ mJ. The harvesting power is set to 0 mW from initial time instant 0 during 7 s and then, 2 mW.

Table 1. Parameters of the jobs

Job J_i	π_i	C_i (s)	r_i (s)	d_i (s)	E_i (mJ)
J_1	1	1	7	13	10
J_2	2	1	5	12	10
J_3	3	1	6	14	2
J_4	4	1	0	15	2

Assume that energy supply is not limited. Figure 5 depicts the FP schedule where all the jobs are executed in the ASAP mode. As no deadline is missing, we say that the jobset is time-schedulable under the FP priority assignment rule with $\pi_1 > \pi_2 > \pi_3 > \pi_4$.

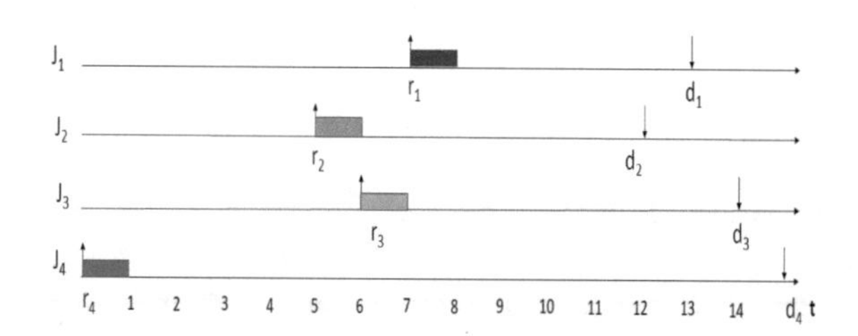

Fig. 5. FP scheduling with no limitation in energy supply

Consider now that we apply a classical scheduler under a fixed priority assignment rule. Figure 6 shows us that, firstly a non idling scheduler cannot face to energy depletion, and second a non clairvoyant scheduler cannot avoid deadline missing caused by energy starvation.

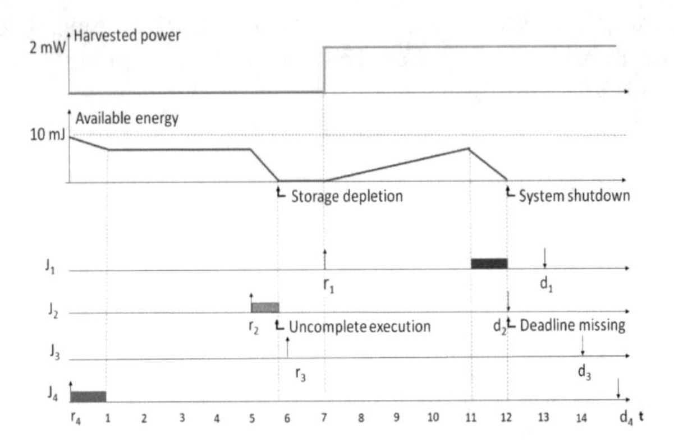

Fig. 6. FP Scheduling with no clairvoyance.

The valid FP-H schedule is depicted on Fig. 7. To decide whether or not J_4 may start execution at $t_c = 0$, the energy availability is checked by computing the preemption slack energy with Eq. (4). As $SE_2(0) = 0$, the processor has to idle so as the energy storage does not deplenish. The slack time is computed so as to determine the latest time when to stop the idle mode according to Eqs. (2) and (3). We have $ST(0) = 10$. However, as the energy storage is full when J_2 releases at time 5, J_2 immediately starts execution since there is no advantage in delaying the job. J_2 finishes at time 6 where the energy storage is totally depleted. Since $ST(6) = 6$, the processor is put in the idle mode until time 12 where J_1, J_2 and J_4 may be executed as late as possible with no deadline missing. Note that J_3 and J_4 execute by consuming energy at the same speed as that it is produced. This example illustrates that the premature consumption of only two units of energy by J_4 was enough to make the system faulty when it was possible to postpone its execution with no deadline violation (see Fig. 6).

5 Implementation Considerations

Probably the greatest difference between energy-neutral and conventional battery-operated systems is their behavior in situations where a system has deplenished its energy storage. At this point, an energy-neutral system switches into a sleep mode and wakes up as soon as enough energy has been harvested to resume task execution. In contrast, a classical battery-operated system reaches the end of its lifetime. For this purpose, an energy-neutral system needs to provide means to safely store its state in persistent memory after having detected

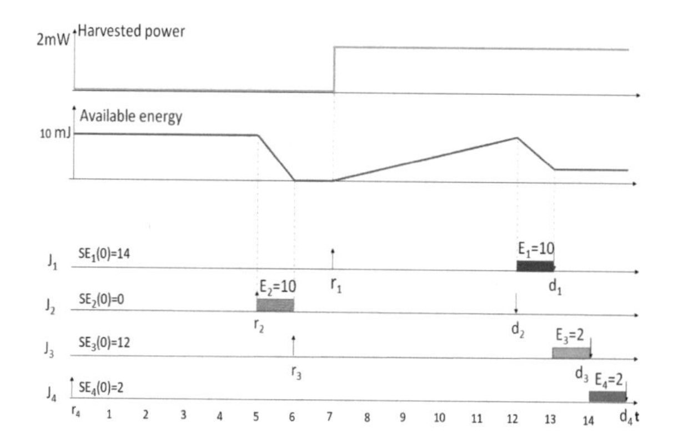

Fig. 7. The optimal FP-H schedule

that no more energy is available in the storage. Furthermore, the system must have detailed knowledge about the energy consumption of tasks, here called WCEC (Worst Case Energy Consumption. In addition, the system must have detailed the minimum energy that should be available in the storage to autorize wake up and guarantee the system to execute tasks for at least a given amount of time.

6 Conclusion

In this paper, we addressed the scheduling issue in a self-powered device with timeliness requirements expressed in terms of deadlines. The central challenge is to make this system energy neutral despite intermittent and variable production of environmental energy used to supply it. The paper reported a new fixed priority based scheduler which smartly defines the busy and sleep periods of the processor. The resulting intermittent computing framework allows to achieve an energy-neutral mode of operation with no wasted energy, no energy starvation and no deadline missing whenever possible. Most RTOS (Real Time Operating Systems) use fixed-priority scheduling where the developers assign each task a suitable static priority level to indicate its relative urgency. Consequently, integration of this new scheduler will permit RTOS to evolve and to meet the rising demand of energy harvesting technology.

References

1. Yildiz, F.: Potential ambient energy harvesting sources and techniques. J. Technol. Stud. **35**(1), 40–48 (2009)
2. Adu-Manu, K.S., Adam, N., Tapparello, C., Ayatollahi, H., Heinzelman, W.: Energy-harvesting wireless sensor networks (EHWSNS): a review. ACM Trans. Sens. Netw. **14**(2) (2018)

3. Ma, D., Lan, G., Hassan, M., Hu, W., Das, S.K.: Optimizing sensing, computing, and communication for energy harvesting IoTs: a survey. IEEE Commun. Surv. Tutor. **22**(2), 1222–1250 (2019)
4. Liu, J.W.S.: Real-Time Systems. Prentice Hall (2000)
5. Davis, R., Cucu-Grosjean, L., Bertogna, M., Burns, A.: A review of priority assignment in real-time systems. J. Syst. Architect. **65**, 64–82 (2016)
6. Liu, C.-L., Layland, J.-W.: Scheduling algorithms for multiprogramming in a hard real-time environment. J. Assoc. Comput. Machin. **20**(1), 46–61 (1973)
7. Allavena, A., Mosse, D.: Scheduling of frame-based embedded systems with rechargeable batteries. In: Workshop on Power Management for Real-Time and Embedded Systems (2001)
8. Moser, C., Brunelli, D., Thiele, L., Benini, L.: Real-time scheduling for energy harvesting sensor nodes. Real-Time Syst. **37**(3), 233–260 (2007)
9. Liu, S. ,Qiu, Q., Wu, Q.: Energy aware dynamic voltage and frequency selection for real-time systems with energy harvesting. In: Proceedings of Design, Automation and Test in Europe, pp. 236–241 (2008)
10. Liu, S., Lu, J., Wu, Q., Qiu, Q.: Harvesting-aware power management for real-time systems with renewable energy. In: IEEE Transactions on Very Large Scale Integration (VLSI) Systems, pp. 1–14 (2011)
11. Abdeddaïm, Y. , Chandarli, Y., Masson, D.: The optimality of PFPASAP algorithm for fixed-priority energy-harvesting real-time systems. In: Proceedings of the Euromicro Conference on Real-Time Systems (ECRTS) (2013)
12. Chetto, M., Queudet, A.: A note on EDF scheduling for real-time energy harvesting systems. IEEE Trans. Comput. **63**(4), 1037–1040 (2014)
13. Chetto, M.: Optimal scheduling for real-time jobs in energy harvesting computing systems. IEEE Trans. Emerg. Top. Comput. **2**(2), 122–133 (2014)
14. Sandhu, M.M., Khalifa, S., Jurdak, R., Portmann, M.: Task scheduling for energy-harvesting-based IoT: a survey and critical analysis. IEEE Internet Things J. **8**(18), 13825–13848 (2021)
15. Chetto, M.: Fixed Priority Scheduling and Energy Neutrality for Autonomous Embedded Real-Time Systems. Technical Report, Nantes Université (2023)
16. Kansal, A., Srivastava, M.B.: An environmental energy harvesting framework for sensor networks. In: Proceedings of the International Symposium on Low Power Electronics and Design (ISLPED 2003), pp. 481–486 (2003)

Energy-Aware Anomaly Detection in Railway Systems

Manuel Mazzara[1] and Alberto Sillitti[2(✉)]

[1] Innopolis University, Innopolis, Russian Federation
m.mazzara@innopolis.ru
[2] Centre for Applied Software Engineering, Genova, Italy
alberto@case-research.it

Abstract. Anomalies in switches behavior in railway systems can significantly impact operational efficiency and safety. This paper proposes an approach based on energy consumption data of switches to identify abnormal behaviors. The approach is based on statistical analysis for real-time anomaly detection in switch energy measurements. Our approach is general enough to be applied to a large number of models of switches, using different technologies, and installed in different environmental conditions. The models presented have been developed using both simulated data and real ones from a large Italian manufacturer.

Keywords: anomaly detection · energy-aware systems · statistical models

1 Introduction

The increasing stress on railway infrastructure, driven by the evolution of mass transportation seeking faster connections and higher transport capacity, requires improved anomaly detection activities to improve safety while maximizing the infrastructure usage. Furthermore, the ability to quickly respond to unexpected events and monitor the entire infrastructure in real-time is crucial. To address these challenges, an advanced system able to monitor and analyzing continuously data is required. One of the main challenges of such systems is the ability to adapt to the different technologies and environments present in complex railways systems.

Our approach focuses on the development of an (almost) agnostic approach to detect anomalies in railway switches based on the continuous statistical analysis of their energy consumption. This paper extends the work performed in the EU-funded project MANTIS[1], emphasizing the integration of energy-aware anomaly detection in the railway system. By incorporating energy consumption data, the approach aims to detect anomalies in real-time requiring a limited amount of previous knowledge about the infrastructure technology and specific deployment.

The paper is organized as follows: Sect. 2 briefly summarizes previous works; Sect. 3 summarizes the main structure of a railway switch; Sect. 4 introduces our approach; finally, Sect. 5 draws some conclusions and introduces future work.

[1] https://www.kdt-ju.europa.eu/projects/mantis.

J. Bravo and G. Urzáiz (Eds.): UCAmI 2023, LNNS 841, pp. 235–242, 2023.
https://doi.org/10.1007/978-3-031-48590-9_23

2 Previous Works

Anomaly detection research has gained significant popularity, particularly in cyber-physical systems, as it plays a crucial role in enhancing operational efficiency and safety. It is widely studied in the most critical domains such as aviation and Industrial Internet of Things [1–3]. A very diverse range of techniques are applied starting from simple statistical analysis to the most complex machine learning and neural networks techniques [4–6].

Moreover, energy-aware anomaly detection is very popular in some domains such as in the smart buildings [7–10]. However, the application of such energy-aware techniques in the railway domain, especially for switches, remains largely unexplored. Most of the work related to the railway infrastructures deals with proactive maintenance techniques and forecasting rather than real-time anomaly detection [11–13].

A deeper analysis of the literature related to the motivation for this work and the current state of the art can be found in [4, 5].

In conclusion, while the significance of anomaly detection research and the application of energy-aware techniques in various domains are well-established, the specific integration of energy-aware anomaly detection in railway switches lacks investigation. This paper contributes to the field by proposing a novel approach for detecting anomalies in railway switches based on the analysis of energy consumption data, offering potential improvements in operational efficiency and safety.

3 Structure of a Railway Switch

A railway switch is a critical component of railway infrastructure that enables trains to change track. It includes several key elements (Fig. 1):

1 **Stock Rail:** it is a long, straight rail that serves as the main track on which the train wheels normally run. It extends from the diverging end of the switch to a point beyond the frog (see point 3).
2 **Switch Rail:** it consists of two curved rails that diverge from the stock rail at the point of the switch. It guides the train wheels onto the desired track when the switch is properly aligned.
3 **Frog:** it is a diamond-shaped component located at the junction of the stock rail and the switch rail. It provides a smooth transition for the train wheels from one track to another. It has a moveable point to allow the alignment of the switch.
4 **Point:** it is the movable section of the switch rail that can be adjusted to align to either the stock rail or the switch rail. It is connected to the switch mechanism and can be moved to direct the train wheels to the desired track.
5 **Switch Mechanism:** it consists of several mechanical components that control the movement of the point. It includes a lever, rods, and connecting mechanisms that allow the point to be moved manually or remotely, depending on the type of switch.

The structure of a railway switch may vary depending on the type and configuration of the switch, as well as the specific railway system requirements. However, the fundamental elements mentioned above form the basis of a typical switch, allowing trains to safely navigate between different tracks.

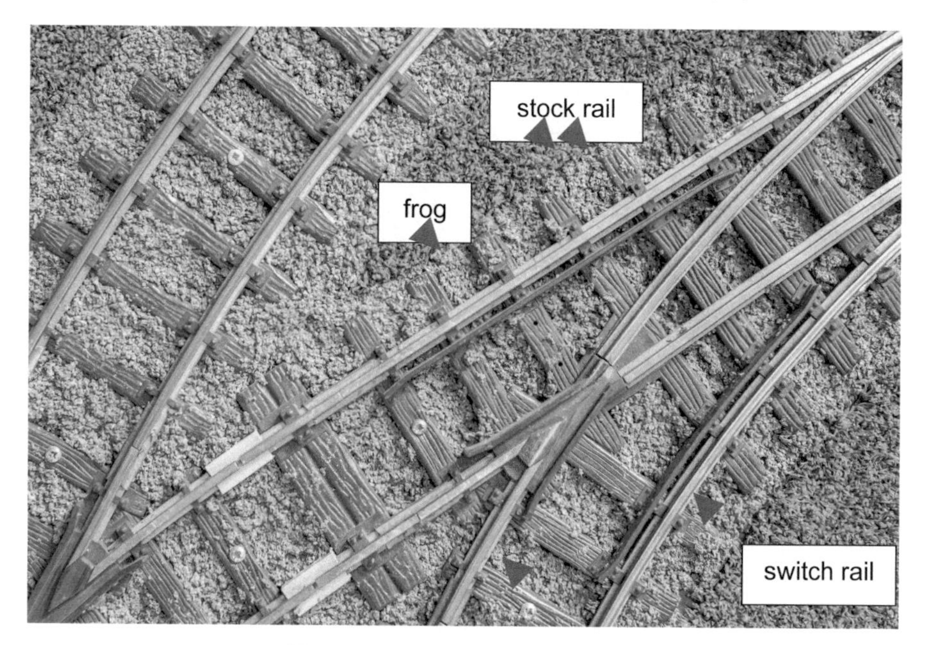

Fig. 1. Main rail switch components.

4 Our Approach

The typical profile of energy consumption of a railway switch is displayed in Fig. 2. The profile is characterized by:

1. **Initial consumption peak:** this is the energy required to start the physical movement of the points. This is the highest level of consumption in the entire movement.
2. **Plateau:** the energy consumption is almost constant over the entire movement.
3. **End of movement:** in the last part of the movement the consumption increases due to the extra effort required to the motor when the points reach the final position. Then, it goes to zero when the movement is completed.

This is the energy consumption of a normal movement. Any significant deviation is an anomaly. However, every switch is different, and the related energy consumption profile is different. Moreover, the environmental conditions affect the energy consumption profile. Therefore, identifying what is normal and what it is abnormal is not straightforward.

As described in [13], we have aligned over time the collected samples to have all of them matching the starting point and we have performed a very basic initial analysis to understand the type of distribution of the data for each sampling time. We applied a Shapiro-Wilk test to verify the non-normality of the distribution (Fig. 3).

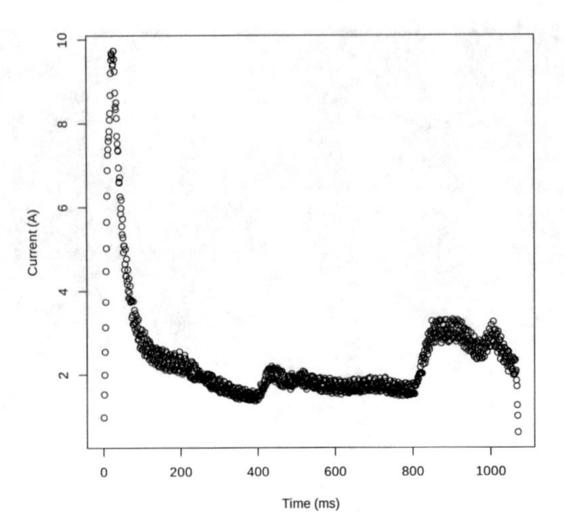

Fig. 2. Typical energy consumption of a switch.

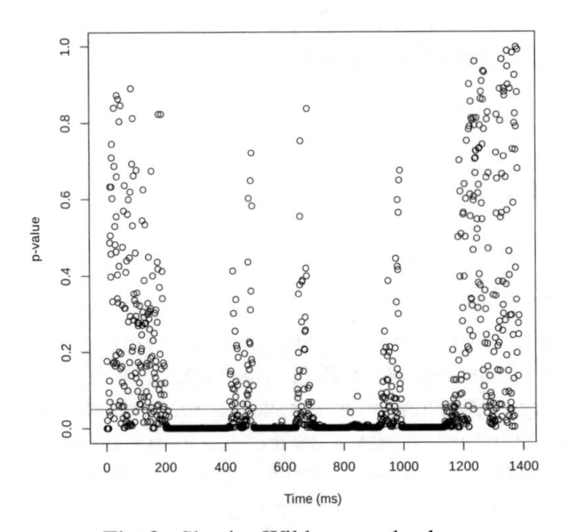

Fig. 3. Shapiro-Wild test on the data.

After that, we investigated the energy consumption peaks in summer and winter (Fig. 4) since we expect that the overall temperature of the environment may affect the movements of the switches. To do that, we analyzed the distribution of the collected samples and we realized that they are normal for both Summer and Winter. Therefore, using a F-test we checked the significance of the difference of the variances and a Wilcoxon test for the medians. The results pointed out that only the difference of medians is significant ($p < 0.01$) showing a similar overall behavior in the entire year regardless the specific season.

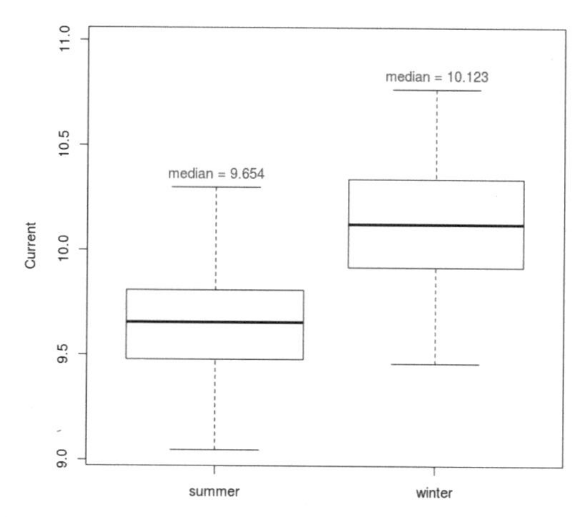

Fig. 4. Comparison of energy peaks in Summer and Winter.

We also investigated the presence of outliers based on the Tukey's range test [14] for outliers [q1–1.5*IRQ, q3 + 1.5*IRQ]. We defined the outliers ranges using a random subset of movements and we applied the bootstrap approach to improve the confidence in the results. Looking at the diagram (Fig. 5), we can state that even with a low number of movements considered to define the outlier boundaries for the peak (about 30), the outliers are less than 1%. Due to this very small number, the calculation of the confidence interval is nor able to converge even with 100,000 replicas.

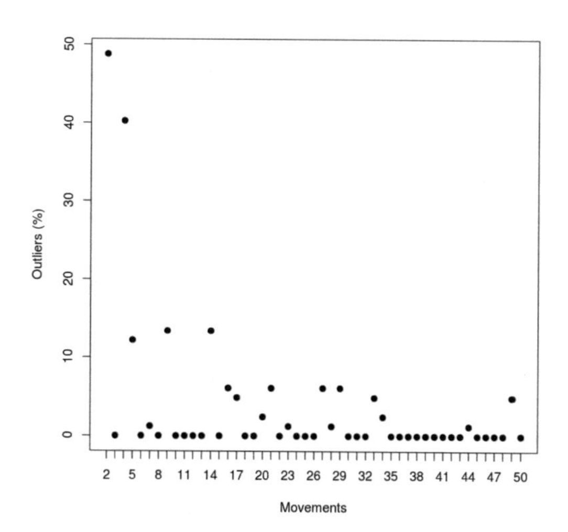

Fig. 5. Analysis of the outliers.

Since there is a significant difference in the data, the different the profiles of energy consumption need to be calculated separately. Figure 6 shows the medians in the Summer and in the Winter to visually understand the differences:

1 More overall energy required in the Winter.
2 Higher consumption peak in the Winter.
3 More time required to complete the movement in the Winter (as already investigated and reported in [13].

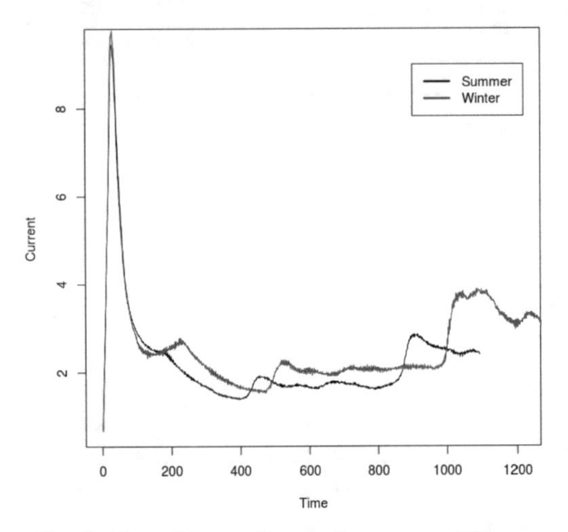

Fig. 6. Plot of the medians in Summer and Winter.

Overall, with a calibration of about 30 movements in the different seasons it is possible to build a model describing the correct behavior of the switch. Since the data required and the computation are limited, this calibration can be easily performed on the field.

About the identification of the abnormal movements using this simple energy-based approach, we plotted some abnormal movements on a diagram displaying the normal movements data (Fig. 7). In most of the cases, such movements are easy identified outside the outliers bands identified with the Tukey's range test.

Based on the results, we have calculated some common performance metrics. The calculated precision is 0.89 and the recall is 0.85.

However, such performances are not based on an adequate amount of data since the dataset used included only a limited amount of real abnormal measurements (<1%). To rebalance the data, we have used mathematical techniques (e.g., super sampling, down sampling, etc.) but the availability of more real data could improve our confidence in the results.

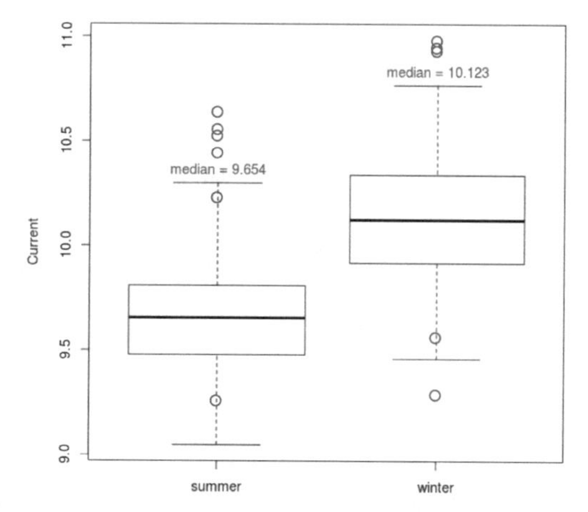

Fig. 7. Abnormal movements identification.

The data used to develop the model are summarized in Table 1 and we have found out that the behavior is almost the same for every model considered even if the technologies used are different (e.g., mechanics, electronics, size, etc.). However, the limited amount of real abnormal data may have affected our evaluation.

Table 1. Summary of the dataset used to develop the model.

Switch model	Number of movements	Abnormal movements
1	1,521	11
2	220	0
3	81	0
4	1,538	18
5	89	0
6	2,597	27
Total	**6,046**	**56**

5 Conclusions and Future Work

The paper has investigated the feasibility of using the energy consumption of a railway switch to detect abnormal behaviors in real time with no knowledge about the actual characteristics of the switches (e.g., technology, operating limits, electrical and mechanical characteristics, etc.). The results are promising but a deeper investigation with more real data is needed to improve the confidence in the reported results.

References

1. Behera, S., Choubey, A., Shambhubhai Kanani, C., Singh Patel, Y., Misra, R., Sillitti, A.: Ensemble trees learning based improved predictive maintenance using IIoT for turbofan engines. In: 34th ACM Symposium on Applied Computing (SAC 2019), Limassol, Cyprus (2019)
2. Behera, S., Misra, S., Sillitti, A.: Multiscale deep bidirectional gated recurrent neural networks based prognostic method for complex non-linear degradation systems. Inf. Sci. 524, Elsevier, 120–144 (2021)
3. Behera S., Misra S., Sillitti A.: GAN-based multi-task learning approach for prognostics and health management of IIoT. Trans. Autom. Sci. Eng. **99**, 1–21 (2023)
4. Albano, M., Jantunen, E., Papa, G., Zurutuza, U.: The MANTIS Book: Cyber Physical System Based Proactive Collaborative Maintenance (Eds.), River Publishers (2019)
5. Fahim, M., Sillitti, A.: Anomaly detection, analysis and prediction techniques in IoT environment: a systematic literature review. IEEE Access **7**, 81664–81681 (2019)
6. Sillitti A., et al.: Providing proactiveness: data analysis techniques portfolios. In: Albano, M., Jantunen, E., Papa, G., Zurutuza, U. (eds.) The MANTIS Book: Cyber Physical System Based Proactive Collaborative Maintenance. River Publishers (2019)
7. Fahim, M., Sillitti, A.: An anomaly detection model for enhancing energy management in smart buildings. In: IEEE International Conference on Communications, Control, and Computing Technologies for Smart Grids (SmartGridComm 2018), Aalborg, Denmark (2018)
8. Fahim, M., Sillitti, A.: Forecasting energy consumption in residential buildings using ARIMA models. In: 6th International Conference on Time Series and Forecasting (ITISE-2019), Granada, Spain (2019)
9. Fahim M., Sillitti A.: Analyzing load profiles of energy consumption to infer household characteristics using smart meters. Energies **12**(5), MDPI (2019)
10. Fahim, M., Fraz, K., Sillitti, A.: TSI: time series to imaging based model for detecting anomalous energy consumption in smart buildings. Inform. Sci. **523**, Elsevier, pp. 1–13 (2020)
11. Hegedűs C., et al.: Proactive maintenance of railway switches. In: 5th International Conference on Control, Decisions and Information Technologies (CoDIT 2018), Thessaloniki, Greece (2018)
12. Socorro, R., et al.: Success stories on real pilots. In: Albano, M., Jantunen, E., Papa, G., Zurutuza, U. (eds.) The MANTIS Book: Cyber Physical System Based Proactive Collaborative Maintenance. River Publishers (2019)
13. Papa, G., et al.: Improving the maintenance of railway switches through proactive approach. Electronics **9**(8), MDPI (2020)
14. Tukey, J.W.: Comparing individual means in the analysis of variance. Biometrics **5**, 99–114 (1949)

Raise Awareness of the Environmental Impacts of Retail Food Products: A User-Centered Scenario-Based Approach

Lorenzo Porcelli and Francesco Palmieri(✉)

Department of Computer Science, University of Salerno, Fisciano, SA, Italy
{lporcelli, fpalmieri}@unisa.it

Abstract. The climate is warming rapidly, and atmospheric concentrations of greenhouse gases (GHGs) are at their highest levels ever recorded. As a result of these climate changes, caused mainly by human activities, disasters have increased fivefold over the past 50 years, causing death and economic loss. Civic engagement and awareness are essential to mitigate climate change and its impacts. In this work, we proposed a user interface that makes users aware of the environmental impact of the food products they buy when shopping. A user-centered scenario-based design was followed in the development of the interface. Gamification elements were added to increase civic participation in climate action.

Keywords: Civic engagement · Climate Change · Food Product Footprint · Gamification · Sustainability · User-Centered Design

1 Introduction

Climate change is one of the most pressing challenges that humanity is currently facing. The last eight years were the warmest on record [16]. The Sustainable Development Goals Report 2022 [1] estimates that 700 million people will be displaced by drought alone by 2030, and about one-third of the world's land area will experience at least moderate drought by 2100. The most significant impact will be poverty and hunger, as millions of people won't have access to fundamental services such as healthcare and education. As well as hindering economic growth, increasing inequality can be a source of international conflicts.

To address the world's most pressing challenges, the United Nations proposed a global call for sustainable action to improve people's lives and preserve the planet identifying 17 Sustainable Development Goals (SDGs)[1]. Among the SDGs, Goal 13 (Climate Action) focuses on actions to combat climate change and its effects. Climate change is, in part, a result of human activities that release greenhouse gases (GHGs). The GHGs are the main contributor to global warming. The primary objective of Climate Action is to strive for attaining net zero global greenhouse gas emissions by the year 2050.

[1] Sustainable Development Goals (SDGs): www.un.org/sustainabledevelopment/.

J. Bravo and G. Urzáiz (Eds.): UCAmI 2023, LNNS 841, pp. 243–254, 2023.
https://doi.org/10.1007/978-3-031-48590-9_24

Carbon footprint is a term used to describe the total amount of greenhouse gases produced. The most common greenhouse gas is carbon dioxide. Carbon footprint also includes other gases such as methane, nitrous oxide and fluorinated gases. By understanding and measuring our carbon footprint, we can identify areas where we can reduce greenhouse gas emissions and hence mitigate the impact on climate change.

Current national emission reduction commitments are insufficient to meet the targets set by the SDGs. An important part of building the public support needed for successful climate change policy is raising public awareness [3]. As at least 25% of greenhouse gas emissions come from food [11], understanding the environmental footprint of retail food products can help consumers make more informed choices and reduce their impact. As consumers feed the food supply system, if they make informed choices and prioritize climate improvement, food producers need to adapt to consumer demands.

We proposed a graphical user interface for an application that informs consumers about the food product footprint they buy in a supermarket. In developing the interface we followed the scenario-based approach of Rosson and Carroll [14]. Gamification elements [9] were also added to increase user engagement.

The remainder of the paper is organized as follows. A review of related literature is presented in the next section. Section 3 provides a comprehensive definition of food product footprints and outlines several methodologies employed for their computation. Section 4 describes the methodology followed to create a prototype graphical user interface for a mobile application. Finally, Sect. 5 encompasses the conclusions drawn from the study and outlines potential avenues for future development.

2 Related Work

A recent review [12] of works that studied the impact of footprint labels on consumers showed that consumers have limited awareness of carbon-related measurements, and the current carbon footprint label system remains ambiguous. However, when redesigned using user-friendly symbols such as traffic light colors, consumers' understanding improves significantly.

However, the future of footprint labels is increasingly dematerialized. CarbonCloud[2] provides a free online database called ClimateHub that allows users to search and find information about the carbon footprint of 10,000 branded food and beverage products available in American grocery stores. Each product listed on their website includes a total emissions tag based on its weight. Consumers can access more detailed information by clicking on the product to learn about the percentage of emissions that come from transportation, packaging, processing, and agricultural practices.

It can sometimes be frustrating to check the footprints of purchased products one by one before shopping. Other proposals increase consumer awareness more

[2] CarbonCloud: https://carboncloud.com.

quickly at the end of shopping. Evocco[3] is a startup that has created a mobile app with the same name that helps users determine the carbon footprint of their food purchases and track the environmental impact of their choices. Users can take a photo of their grocery receipt, and the app's machine learning technology identifies the products and calculates their climate impact based on type, weight, and origin.

Our proposal aims to raise user awareness in the most direct way possible. We provide a tool as simple as possible and usable by anyone, which shows the footprint of a food product at the time that matters, i.e., when the product is about to be placed in the shopping basket.

3 Food Product Footprints

Consumers are rarely shown the environmental consequences of producing and consuming food. The environmental footprint of product items allows consumers to compare the impact on the environment between different product groups.

The comprehensive environmental impact of producing a food product must take into account at least sustainability metrics such as carbon, nitrogen and water footprints [5]. The carbon footprint of a product or service reflects the amount of greenhouse gas emissions released throughout its lifecycle, typically including production, use or consumption, and disposal [13]. The nitrogen footprint of a food product indicates the overall quantity of reactive nitrogen released into the environment as a result of the production and consumption of that specific food product [6]. The water footprint of a food product can be defined as the quantity of water used, both through evaporation and transpiration, during the production process of that particular product [4]. The work [5] presents three calculation methods for food product footprints.

The first calculation method, known as the footprint weight, directly shows the weight of the environmental impact of a specific product. Footprint weight is the most commonly used method when determining the environmental impact of a product. It is calculated as

$$F_w = w \times f, \tag{1}$$

where F_w is the footprint calculation based on weight, w is the weight of the product, and f is the footprint factor.

The second calculation method measures the sustainability of a product based on the production process. It looks at specific sustainability measures, such as crop rotation, riparian buffers, and rotational grazing. When a producer meets more of these measures, the sustainability rating of the product increases. In the following formula, the percent of possible sustainability measures F_s for a product is the sum of sustainability measures that apply at a given farm divided by the total possible sustainability measures, i.e.,

$$F_s = \frac{\sum \textit{sustainability measures at farm}}{\sum \textit{all possible sustainability measures}}. \tag{2}$$

[3] Evocco: https://linktr.ee/evocco.

The third method of calculating footprints, called % Daily Value (DV), shows the percentage of a person's total daily footprint attributed to the consumption of a particular product. The % DV is determined based on a reference value representing a sustainable daily environmental impact. The daily allotment of a healthy diet is another way of looking at the footprint of a single food product. The footprint as % DV value can be written as

$$F_{DV} = \frac{F_w}{D_w}, \quad (3)$$

where F_w is the footprint of a product by weight, and D_w is the total daily footprint associated with a healthy diet.

Each of these three footprint calculation methods can be represented graphically. A simple method proposed by [5] is a star rating system that combines the three footprints into a single sustainability measure as an average of the three. The resulting star rating ranges from 0 stars (indicating the least sustainable) to 3 stars (indicating the most sustainable).

4 Materials and Methods

Our main objective was developing an effective interface for an app to increase awareness and civic participation in sustainability while shopping in available markets. We used a user-centric design paradigm to study the potential user base and consider it in all the conceptual design activities. A scenario-based process allowed us identifying the main actors involved and analyze their behavior.

4.1 Consumer Behavior when Shopping

Recent work by [7], which collected data on 144 participants via eye tracking, confirmed the findings of previous studies regarding consumer behavior. Most consumers do not compare available products before buying but purchase the usual product without giving it much thought. Specifically, the average time consumers hold a product in their hands is less than the time needed to read the product information label. For those who compare products before purchasing, price stands out as one of the most important considerations. Based on consumer behavior, we defined two personas described in Table 1.

4.2 Scenarios of Current Practices

We hypothesized two scenarios, described in Table 2, for defining the basic requirements and designing a potential solution.

Considering the most common current practice scenarios, the following claims emerge. Consumers unaware of the footprint in their purchasing decision-making process evaluate products just on convenience and personal preferences. Being aware of their environmental impact, consumers could introduce an additional element of discrimination on a par with the previous conditions. Instead, those

Table 1. Personas

(1) Maria is a 45-year-old housewife with a high school diploma. She goes shopping 2–3 times a week, usually in the morning between errands while her children are at school. As a mother of two children aged 8 and 10, she juggles many responsibilities, including managing the household and her children's schedules. Maria mainly chooses products by brand and convenience. She occasionally jots some things down to buy on a piece of paper, but most of her shopping is habitual. She primarily uses the smartphone to talk to her husband and children
(2) Olivia is a 30-year-old woman with a degree in economics working as an accountant in a large company. She usually prefers shopping alone, either in the evening or at weekends. As a single professional, she has a busy schedule and values her free time. Olivia is tech-savvy and uses her mobile phone for both work and leisure. She is health conscious and buys organic and local produce whenever possible. She is organized and prepares her shopping list in advance using a checklist on her phone

Table 2. Scenarios of current practices

(1) Maria gets up early and gets her children ready for school. She drops them off and goes shopping before picking them up in a few hours. As she enters, she is bombarded with several choices and options, but she has little time to waste. She navigates the store hastily, taking items from the shelves. She wants to save money and make sure she can get everything on her list before she goes to pick up her children from school. Having found everything, she goes to the checkout and leaves the store
(2) It is Saturday morning, and Olivia checks her pantry to see what she needs to buy. She prepares her shopping list using a checklist app and heads toward the supermarket. On arrival at the supermarket, Olivia skims the aisles, comparing prices and reading labels to ensure the products meet her standards. As a health-conscious person, Olivia prefers organic products and checks the country of origin. She also enjoys discovering new and innovative products to recommend to her friends. She fills her basket with healthy and environmentally friendly products and leaves the store satisfied with her choices

who have little time to browse products can benefit from the advice of others who also assess the environmental impacts of a product before making a purchase. Finally, it is worth noting that everyone has a smartphone at their disposal, which they use for various purposes.

We can formally describe the behavior of Maria and Olivia using a Markov chain, with the states outlined in Table 3. In a Markov chain, the probability of each event depends solely on the immediately preceding event. Each consumer transitions between states with varying probabilities based on their behavior. For a generic transition matrix M, we denote the probability of transitioning from state i to state j as $M_{(i)(j)}$. The transition matrices P and Q, which represent the behavior of Maria and Olivia in the current scenarios, are depicted as transition diagrams in Fig. 1a and Fig. 1b, respectively.

Table 3. States in the Markov Chain

Macro-state	State	Description
Preparation	S1	Prepares the shopping list
	S2	Does not prepare a shopping list
Support	S3	Prepares the shopping list using the application with the proposed interface
	S4	Prepares the shopping list with a generic checklist
	S5	Prepares the shopping list with pen and paper
Influence on purchases	S6	Exposure to recommendations of low environmental impact products by other users
	S7	Habitual purchases
	S8	Other influences
Purchase attention	S9	Reads product labels
	S10	Does not read product labels
Item comparison	S11	Compares products based on price
	S12	Compares products based on environmental impact
	S13	Compares products based on other characteristics
Sharing	S14	Shares recommendations on low environmental impact products
	S15	Does not share recommendations on low environmental impact products

4.3 Application Requirements

Scenarios of current practices allowed us to produce the first set of user interface requirements. We describe them below, classifying them into five categories according to Preece et al. [10].

Functional Requirements

- The application allows users to manage a grocery list. This requirement satisfies the general need to store items to be purchased.
- The application allows users to cross out a list item by framing a product identification code (e.g., a barcode or QR-code). This requirement allows users to discover directly the environmental impact of the products they purchase.
- When a product with a high environmental impact is scanned, the application suggests a similar product with a lower environmental impact. This requirement raises awareness of the need to choose low-impact products.
- The application allows users to share information about their purchases with the community. This requirement promotes civic engagement in tackling climate change.

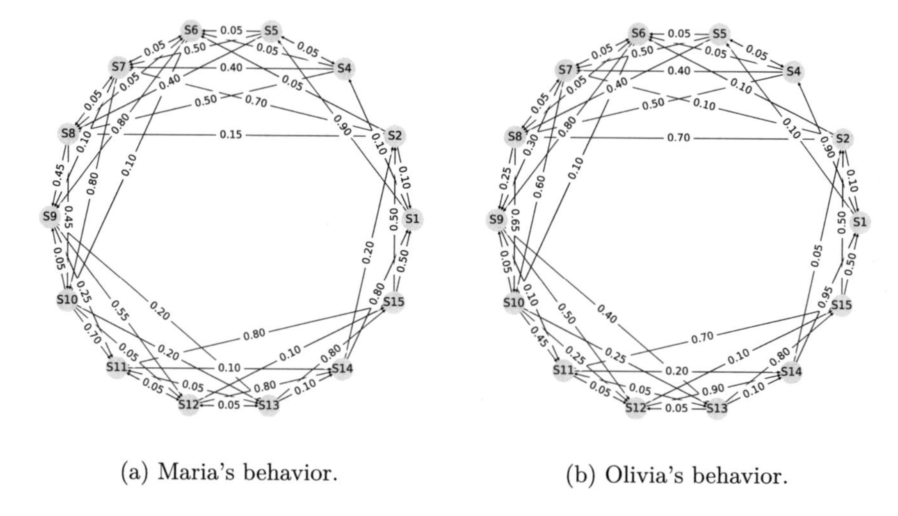

(a) Maria's behavior. (b) Olivia's behavior.

Fig. 1. State transition diagrams of current practices.

Data Requirements

- The application must be able to handle large amounts of data efficiently. This requirement allows user data to be collected, processed and shared without slowing down or crashing.
- The application must be able to manage data synchronization between users. This requirement allows users to share up-to-date information in real-time.

Environmental Requirements

- The application must be compatible with the most popular mobile devices. This requirement ensures broad distribution and use by a large segment of users.
- The application must be usable in environments with different lighting conditions. This requirement ensures visibility and readability in different environmental conditions, including those with low lighting.

User Requirements

- Users already use a smartphone for basic tasks.
- The majority of users are aged between 18 and 60.

Usability Requirements

- The application must be sufficiently easy to use by requiring no specific training effort. This requirement ensures that the app is usable by anyone with a minimum effort.
- The user interface should provide simple management of user mistakes. This requirement reduces frustration due to unintentional user errors.

– The application must be usable by people with different physical abilities. In particular, the application should be accessible to visually impaired users by supporting features such as text-to-speech or operating system font enlargement. This requirement ensures access for all citizens without discrimination.

4.4 Gamification Elements

Gamification involves incorporating game elements into non-game contexts to incentivize the use of a system by tapping into the naturally rewarding aspects of games [2]. We introduced gamification elements into the application, such as points, badges, levels and leaderboard, to help motivate and engage users.

A distinction between intrinsic and extrinsic motivation is well-known in the literature [8]. Intrinsic motivation is the most authentic and self-directed form of motivation, where individuals engage in an activity for the sheer pleasure and personal interest it arouses because they find the activity intrinsically rewarding. On the other hand, extrinsic motivation is based on external rewards, such as a reward or money or the avoidance of a negative consequence.

The Self-determination Theory (SDT) [15] posits that the intrinsic motivation of individuals is influenced by three fundamental psychological needs: Competence, Relatedness, and Autonomy. Competence involves the acquisition of skills to deal effectively with the external environment, including tasks such as solving challenging problems. Relatedness refers to the significance of social connections, encompassing interactions and competition with others. Lastly, Autonomy represents an intrinsic desire to have control over one's life and to act in accordance with own values.

We mapped the gamified elements of the app and their motivation to the three psychological needs defined in Self-Determination Theory (SDT), as shown in Table 4.

Table 4. Gamified features of the application

Feature	Rationale	SDT
Points, Levels	Users earn points scanning products. Users' awareness levels up based on the points they earn	Competence
Mission	A mission has a specific objective, e.g., identifying five products with a smaller environmental footprint in the soft drinks category	Competence, Autonomy
Badges	When users complete a specific mission they earn a badge. Badges are displayed on the user's profile	Competence, Relatedness
Leaderboard, User profile	Users can share their progress and grocery list items with other players	Relatedness

Table 5. Activity transformation scenarios

(1) Maria, knowing that she can find suggestions for quickly creating a grocery list, decides to try the application with the proposed interface. Now she can see the environmental footprint of each product as she puts items into her basket. When Maria has to choose between two products of the same price, she chooses the one with the lowest food footprint
(2) Olivia, a conscientious shopper, consistently ranks at the top of the leaderboard for users who make environmentally conscious purchasing decisions based on the food footprint of each product. Olivia publicly shares the choices that have allowed her to rank at the top of the leaderboard with other users

4.5 Scenarios Transformation

An app that assists users with grocery lists may help consumers make more environmentally conscious choices. Since all consumers are familiar with smartphones for other tasks, the development of a mobile application seems to be an appropriate direction. The idea behind the application has led to a transformation of the scenario reported in Table 5.

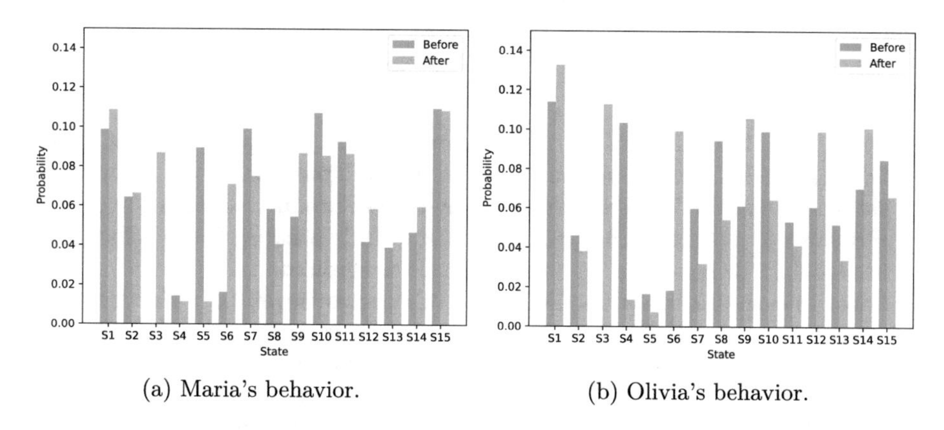

(a) Maria's behavior. (b) Olivia's behavior.

Fig. 2. Stationary distribution of the Markov chain before and after adoption of the application with the proposed interface.

As for the Markov analysis process, the transition diagrams depicted in Fig. 1 undergo an evolution through the inclusion of the state S3, which is described in Table 3. Specifically, in Fig. 1b, we have updated the following transitions: $P_{(S1)(S3)} = 0.80$, $P_{(S1)(S5)} = 0.10$, $P_{(S3)(S6)} = 0.70$, $P_{(S3)(S7)} = 0.15$, and $P_{(S3)(S8)} = 0.15$. In Fig. 1b, the updates encompass the transitions $Q_{(S1)(S3)} = 0.85$, $Q_{(S1)(S4)} = 0.10$, $Q_{(S1)(S5)} = 0.05$, $Q_{(S3)(S6)} = 0.80$, $Q_{(S3)(S7)} = 0.10$, $Q_{(S3)(S8)} = 0.10$, $Q_{(S9)(S12)} = 0.75$, and $Q_{(S9)(S13)} = 0.15$.

By calculating the stationary distribution through numerical experiments, we can analyze the changes in user behavior resulting from the adoption of the application. From the graphs in Fig. 2, it is evident that, in general, the likelihood of being in states S6, S9, S12, and S14 has increased. Specifically, Maria tends to be more careful when shopping. Reading labels and comparing products makes her aware of the environmental impact of her purchases. Olivia, on the other hand, now has a platform where she can easily share her green choices. Olivia knows that, in her own small way, her contributions can positively influence the choices of other users, including Maria.

Based on the described activity scenarios, we identified the most significant design requirements for the application.

- The presentation of the shopping list must be simple and immediate, leveraging the communicative power of images and the language of colors. This allows the small mobile device to provide users with composed and complex information.
- The interface must provide users with a reduced number of tabs for navigating between screens. Users use this application frequently and require rapid usage.
- Checking products at the time of purchase must occur quickly and immediately, as should the display of similar products.

4.6 Designing the Mobile User Interface

Based on the scenarios, claims, and application requirements, we have developed a set of interface requirements to guide the design of our application.

- **Interface that Minimizes Text Input**. As the user starts typing, product suggestions containing the substring of the input text based on previous shopping lists are displayed. Frequent and latest products are also suggested.
- **Product List with Images**. Users can quickly identify the product they are looking for regardless of their cultural background.
- **Interface that Limits the Number of Interactions**. Only the necessary elements for completing a task are shown, while everything else is hidden.
- **Identification Code Scanner**. Users can quickly select purchased items by scanning the barcode or QR code available on its packaging.
- **Product Footprint Label**. After the user scans the identification code of a product, a star rating system [5] with the footprint is shown.

In terms of the application's information architecture, when users enter the application, they can create a new shopping list, and suggestions are made based on their previous lists to help them create a new one. Suggestions may include low-carbon alternatives based on the purchases of other users.

Once a grocery list is created, users can view it and mark the products added to their basket by scanning the relevant identification code. As each product is scanned, its food product footprint is displayed, increasing the user's awareness of its impact on the environment. Figure 3 shows the application screens relating to a shopping list, the scanning of a product, and the crossing out of a product to be added to the basket.

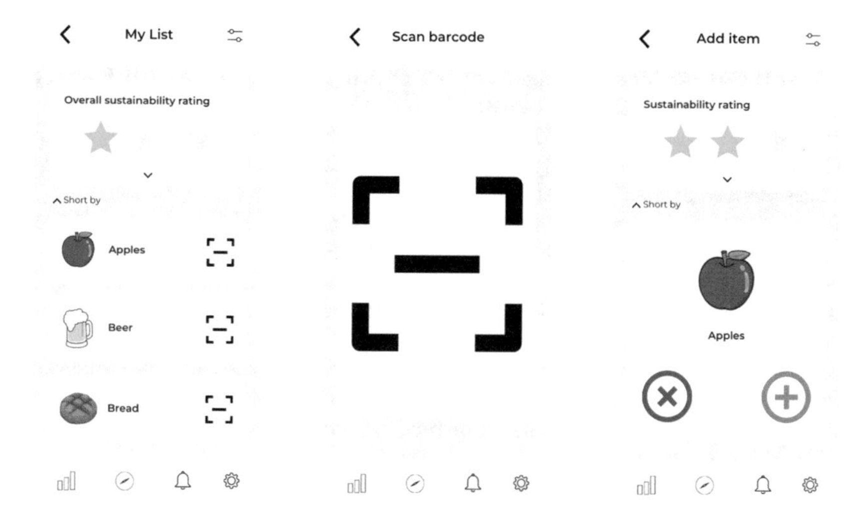

Fig. 3. Application User Interface prototype

5 Discussion and Conclusions

The development of mobile applications is an effective way to increase user awareness and engagement in activities of common interest. One of the main challenges is to create a user-friendly and usable interface for all types of users.

The design requirement that guided our work is to make consumers aware of their impact on climate change directly and tangibly. The artifact is a mobile application that enables users to generate a shopping list while assigning an estimated food product footprint and allows users to be aware of other products similar to the ones they use but with a lower footprint.

The application prototype has been designed using user-centered design principles. The scenario analysis helped to identify the different actors involved and how they interact with the products during the purchase process.

The next step in this research will involve conducting a field usability test with customers in a supermarket to demonstrate the effectiveness of the initial solution. Further evaluation and testing are necessary to determine the effectiveness of the artifact in achieving its intended goals and to identify any areas for improvement.

References

1. Department of Economic and Social Affairs: The Sustainable Development Goals. Technical Report, United Nations (2022)
2. Deterding, S., Dixon, D., Khaled, R., Nacke, L.: From game design elements to gamefulness: defining "gamification". In: Proceedings of the 15th International Academic MindTrek Conference: Envisioning Future Media Environments, pp. 9–15 (2011)

3. Drummond, A., Hall, L.C., Sauer, J.D., Palmer, M.A.: Is public awareness and perceived threat of climate change associated with governmental mitigation targets? Clim. Change **149**, 159–171 (2018)
4. Hoekstra, A.Y., Mekonnen, M.M., Chapagain, A.K., Mathews, R.E., Richter, B.D.: Global monthly water scarcity: blue water footprints versus blue water availability. PLoS ONE **7**(2), e32688 (2012)
5. Leach, A.M., et al.: Environmental impact food labels combining carbon, nitrogen, and water footprints. Food Policy **61**, 213–223 (2016)
6. Leach, A.M., Galloway, J.N., Bleeker, A., Erisman, J.W., Kohn, R., Kitzes, J.: A nitrogen footprint model to help consumers understand their role in nitrogen losses to the environment. Environ. Develop. **1**(1), 40–66 (2012)
7. Machín, L., et al.: The habitual nature of food purchases at the supermarket: implications for policy making. Appetite **155**, 104844 (2020)
8. Malone, T.W.: Toward a theory of intrinsically motivating instruction. Cogn. Sci. **5**(4), 333–369 (1981)
9. Pelling, N.: The (short) prehistory of gamification, funding startups (& other impossibilities). J. Nano Dome (2011)
10. Preece, J., Sharp, H., Rogers, Y.: Interaction design. Apogeo Editore (2004)
11. Ritchie, H., Rosado, P., Roser, M.: Environmental impacts of food production. Our World in Data (2022). https://ourworldindata.org/environmental-impacts-of-food
12. Rondoni, A., Grasso, S.: Consumers behaviour towards carbon footprint labels on food: a review of the literature and discussion of industry implications. J. Clean. Prod. **301**, 127031 (2021)
13. Röös, E., Sundberg, C., Hansson, P.A.: Carbon footprint of food products. Assessm. Carbon Footprint Diff. Indust. Sect. **1**, 85–112 (2014)
14. Rosson, M.B., Carroll, J.M.: Scenario-Based Design. L. Erlbaum Associates Inc., USA (2002)
15. Ryan, R.M., Deci, E.L.: Self-determination theory and the facilitation of intrinsic motivation, social development, and well-being. Am. Psychol. **55**(1), 68 (2000)
16. World Meteorological Organization (WMO). Past eight years confirmed to be the eight warmest on record. https://public.wmo.int/en/media/press-release/past-eight-years-confirmed-be-eight-warmest-record. Accessed 23 July 2023

Author Index

J. Bravo and G. Urzáiz (Eds.): UCAmI 2023, LNNS 841, pp. 255–256, 2023.
https://doi.org/10.1007/978-3-031-48590-9

Printed in the United States
by Baker & Taylor Publisher Services